手 绘 名 物 系 列
Hand - Drawn Classic Travel Landmarks

中国民居
Traditional Chinese Residences

王其钧 著
Written and Illustrated by Wang Qijun

刘海乐 译
Translated by Liu Hailang

China Pictorial Press · Beijing

图书在版编目（CIP）数据

中国民居：汉英对照 / 王其钧著. -- 北京：中国画报出版社，2023.4
（手绘名物系列）
ISBN 978-7-5146-2078-8

Ⅰ. ①中… Ⅱ. ①王… Ⅲ. ①民居—建筑艺术—中国—汉、英 Ⅳ. ①TU241.5

中国版本图书馆CIP数据核字(2022)第042311号

中国民居（汉英对照）

王其钧 著

出 版 人：方允仲
项目主持：方允仲　齐丽华
责任编辑：刘晓雪
英文翻译：刘海乐
美术设计：赵艳超
责任印制：焦　洋

出版发行：中国画报出版社
地　　址：中国北京市海淀区车公庄西路33号　邮编：100048
发 行 部：010-88417438　010-68414683（传真）
总编室兼传真：010-88417359　版权部：010-88417359

开　　本：16开（787mm×1092mm）
印　　张：11.75
字　　数：150千字
版　　次：2023年4月第1版　　2023年4月第1次印刷
印　　刷：北京汇瑞嘉合文化发展有限公司
书　　号：ISBN 978-7-5146-2078-8
定　　价：118.00元

前言

本书是“大美中国”书系中的一本。顾名思义，所谓“大美”，无外乎自然界和人类社会中，具有高度审美价值的客体、对象。它们可以是自然界中的山川河流，可以是鸟兽虫鱼，也可能是风雨雷电、莺歌燕舞，更有可能是通过色彩、线条、声音、文字等人文艺术表现出来的各种艺术形式，如音乐、舞蹈、建筑、书法、绘画、雕塑、戏剧、电影等。而这些，都来源于东方古老的文明大国——中国。

浑浊而奔腾不息的黄河孕育了古老的华夏文明。中国人的先祖从母亲河走来，由部落壮大成为部落国家，从王国跨越到帝国，经历了漫长的封建时代。最终，在近代的数次实践之后，中国走上了社会主义新中国的道路——在这漫长的历史中，有统一，也有分裂；有强盛，也有衰败。部落演进为国家的历史，中国经历了一千多年，封土建国的周朝，前后存在了八百多年，之后漫长的封建时期，历时两千余年才告结束——可中国的历史，却又不止于长：不仅长，还富于变化。中国五千多年的文明和国家史，从来不是一成不变的：在四千余年的建国史中，中国经历了约十五个朝代，六七十个政权交替更迭，更见证了不计其数的大小战争——战争和政权更替固然给人民带来了深重的灾难，却又在客观上催动着社会生活方方面面的变化，为中国展开了一幅波澜壮阔的历史图卷；从天文、历算到农耕器具的进步，从造纸、印刷到选官制度的变化、商业的繁荣，中华民族就这样一步步走来，走到今天。

Foreword

This book is part of the "Beautiful China" book series. As its name implies, the book series focuses on depicting objects of high aesthetical value in nature and human society. They include not only creatures such as birds, beasts, insects and fish but also natural phenomena such as wind, rain, thunder and lightning. They may also be various forms of art composed with colors, lines, sound and texts, such as music, dance, architecture, calligraphy, painting, sculpture, drama, and movie. Regardless of their form, all of them derive from China, an ancient civilization in the East.

The muddy, ceaseless flow of the Yellow River fostered the ancient Chinese civilization. Nourished by the "mother river," Chinese ancestors gradually developed their tribes into tribal nations, then kingdoms and empires across the feudal era. Eventually, after several attempts of practical exploration in modern times, China embarked on a path of socialism with Chinese characteristics. Throughout its long history, China experienced unifications and separations as well as ups and downs. The history of tribal nations in China lasted for more than 1,000 years, followed by the Zhou Dynasty (1046-256 BC) that spanned about 800 years. Then, China entered the feudal period as long as more than 2,000 years. However, the course of China's history is more than just long; it is also full of changes. Across its history of more than 5,000 years, the Chinese civilization has never stopped the pace of evolution. Throughout its history of more than 4,000 years as a nation state, China fostered 15 dynasties and some 70 regimes, and underwent numerous wars. Of course, wars and regime shifts might bring grave disasters to the people, but at the same time they accelerated changes in all aspects of society and composed the eventful picture of Chinese history. From progress in astronomy, calendrical science and farming tools, and improvements in printing and papermaking techniques

习近平主席曾多次指出，“当今世界正经历百年未有之大变局”；正是在这样的历史、时代机遇和文化背景下，中国外文局责成中国画报出版社推出了“大美中国”书系。本书系既是对中国自然、风物、人文的总结，也是向全世界关心中国、热爱中国、崇敬中国的人，讲好中国故事。

但凡讲故事，必要有一个主题，有一个入口；我们选择向世界展现中国之美，这并非任意为之。人们总是喜欢美好的东西，这非但是中国人民的追求，也是全世界各国人民、各个民族的追求——不同文化对美的定义或许不同，可这追求却是一致的，而且古来有之。

中国之美既有历史的深邃厚重，又有地域的辽远广阔。大到富丽繁华的都市，小到鲜有问津的村镇，历史总会在这里那里，留下星星点点的痕迹。地域的辽阔又为中国之美赋予了不同的风格。我们既有高山大川，又有小桥流水；既有高屋广厦，又有陋室闲庭；既有“长河落日圆”的雄壮，又有“清泉石上流”的清丽……疆土有多少寸，美就有多少种，简直是说也说不尽的。以自然地理和城镇乡村为依托，我们更想展现一种人文的厚度，南北东西、平原山地、沿海内陆……不同地方的人，他们都在以什么样的方式生活着，他们的生活有何不同，而这种种生活又如何共同构成了中国的文化和民生的一部分。一言以蔽之，我们想要传达的，不是某种抽象的概念，不是简单的意象或标准化的符号，而是由真实、具体的细节所构成的，鲜活的生活与生命。

本书系通过文图作品讲述中国故事；以画面语言为主，辅以文字的叙述、

and official selection systems to prosperity of commerce, the Chinese nation have constantly moved forward step by step to the present day.

Chinese President Xi Jinping said on many occasions that the world is undergoing "profound changes unseen in a century". Facing such historic opportunities and cultural context, under the instruction of China International Publishing Group, China Pictorial Press presented the "Beautiful China" book series, which not only reviews the natural and cultural sights of China but also tells China's stories to global readers who care, love and admire China.

A story needs a theme or topic. We didn't casually choose "Beautiful China" as the theme of the stories we tried to tell. All people love beautiful things. In fact, the pursuit of beauty represents the common aspiration of not only the Chinese people but also people from other parts of the world. The definition of beauty may differ for different cultures, but people around the world have had a shared aspiration for beauty since ancient times.

The beauty of China stems from its profound history and vast territory. Whether in spectacular, prosperous metropolises or in nondescript small towns, we can always find clues left by history. The diverse cultures scattered around the country's vast territory bestow on China different types of beauty. There are high mountains and big rivers as well as exquisite bridges and murmuring streams; there are skyscraping buildings as well as simple dwellings; there are magnificent views of torrential rivers in the setting sun as well as elegant sights of clear streams flowing through rocks... The beauty of China is just as boundless as its territory. With this book, we would like to show readers the natural geography and cultural profundity of various cities, towns and villages across China, from plains to mountains and from coastal areas to inland areas, as well as how people in different regions live in

解释和说明，以给读者更完善的印象和更系统的知识。

除了摄影作品，我们还将绘画艺术融入到书系当中。例如“手绘名物系列”，我们选择了通过水彩画的方式去讲述城市及其民俗的故事，这不仅构成了对城市的刻画，同时也是一次上好的艺术和美学教育。选择水彩画这一既为亚洲也为欧美所熟悉的艺术表现形式，去讲述城市这一最为大多数人所熟悉的生活空间，我们想以此作为出发点，走近听众，把中国故事讲好，把故事讲得生动真实、可闻可感。

沿着这样的轨迹，我们希望把中国最美的一面展示给世界，也想把中国的故事讲给全世界每一个喜欢她的人听。

different ways and how their lives together compose the country's diverse cultures and lifestyles. In one word, we intend to reveal the vigor of life through true and specific details, rather than an abstract concept, a simple imagery or a standardized symbol.

This book series tells stories about China through vivid pictures and Langnage. Therefore, it prioritizes images, supplemented with textural narrations, explanations and remarks, so as to enable readers to obtain deeper impressions and systematic knowledge.

In addition to photographic works, we also incorporate painting into the book series. For example, the "Hand-Drawn Classic Travel Landmarks" sub-series feature hand-drawn watercolor illustrations depicting cities and their folk customs. This is not only an ideal way to portray cities, but also provides a chance for art and aesthetic education. Watercolor is a form of art familiar to both Asian and Western audiences. We chose this form of art to portray cities, a kind of living spaces familiar to most people, in an effort to tell China's stories in a vivid, perceptible way and make them closer to readers.

By doing so, we hope to show the most beautiful side of China to the world and tell China's stories to every reader who is interested in the country and its culture.

Contents

前言
Foreword

概述
Introduction 001

上篇 Chapter I

形式集合
Various Forms of Residences

01 蒙古包
Mongolian Yurts 024

02 藏族碉房
Tibetan Folk Houses 027

03 窑洞
Cave Dwellings 032

04 朝鲜族民居
Traditional Korean Houses 036

05 东北民居
Traditional Residences in Northeastern China 038

06 北京四合院
Siheyuan in Beijing 040

07 江浙民居
Traditional Residences in Jiangsu and Zhejiang 044

08 四川民居
Traditional Sichuan Residences 054

09 云南一颗印民居
Seal-like Compounds in Yunnan 060

10 长方形土楼
Rectangular Tulou 062

11 环形土楼
Circular Tulou 066

12 福建民居
Traditional Residences in Fujian 070

13 广东民居
Traditional Residences in Guangdong 074

14 湘鄂赣民居
Traditional Residences in Hunan, Hubei, and Jiangxi 076

15 皖南民居
Traditional Residences in Southern Anhui 084

16 晋陕民居
Traditional Residences in Shanxi and Shaanxi 092

17 干栏式民居
Stilt Houses 098

18 维吾尔族民居
Uygur Residences 106
19 贵州石板房
Stone Houses in Guizhou 114
20 井干式民居
Log Houses 116
21 白族民居
Bai-style Houses 118
22 西北民居
Traditional Residences in Northwestern China 120

下篇 Chapter 2

村镇面貌
Villages and Towns

01 村镇选址
Site Selection for Villages and Towns 124
02 街道尺度
Streets and Lanes 130
03 死胡同
Dead Ends 132
04 广场
Public Squares 134
05 变幻的街景
Changing Streetscapes 142
06 牌坊
Memorial Archways 150
07 河道
River Courses 152
08 桥
Traditional Bridges 154
09 路棚
Roadside Booths 161
10 私用码头
Private Quays 162
11 祠堂公馆
Ancestral Temples and Guildhalls 164
12 公用空间
Public Spaces 168
13 村镇举例——罗城古镇
Luocheng—A Good Example of Ancient Towns 170
14 环境
Surrounding Environment 174

概述

Introduction

晨曦中，我们徜徉在老村古镇间，穿梭于深巷高墙下，呼吸着清新自由的空气，远离大都市喧闹的嘈杂声。那雄峻的高山，那平静的小河，弥漫着轻烟般的晓雾。一幢幢有形貌有神灵的民居在朝霞下醒来，触动了我们的想象、情感和审美感受。

在福建北部山区的一组民居里（见图一），我们可以看到变化丰富的封火

图一 福建北部山区的一组民居
Fig. 1 A group of traditional residences in a mountainous area of northern Fujian Province.

Bathed in the warm light of the sunrise, several friends and I roamed the lanes and breathed the fresh air of an old rural town to escape the hustle and bustle of the city. Magnificent mountains and bubbling streams were shrouded in the haze. Amid the idyllic ambience, the residences in the village seemed to awake, inspiring our imaginations, emotions, and aesthetic ideals.

The small town (see Fig. 1) tucked away in the northern mountainous region of Fujian Province is home to buildings in various styles including traditional residences with gables in diverse shapes and neatly tiled roofs, wooden houses projecting over the ground, and small and simple sheds. Serenaded by rooster calls, the town embraced a new day.

Whether traveling in an ancient town in southern Anhui Province or an old village in Guizhou Province, we were frequently spellbound by the beauty of the residential buildings. From the aesthetic perspective, each residence has unique beauty, even for those in the same architectural style. Different residences manifest varying aesthetic appeals and please the eye in different ways.

This residence (see Fig. 2) in southern Anhui features the style of Ming Dynasty (1368-1644) residential buildings. It arouses a sense of solemnity and primitive simplicity. Its overlapped eaves create a dynamic beauty, while its horizontal outlines seem to foster peace and tranquility. The scene hearkened to the lines by Tang Dynasty (618-907) poet Wen Tingyun in *Setting out Early from Mount Shang*: "The cry of the rooster, the moon over a thatched inn, and someone's tracks frosted over on the plank bridge." The poetic scenery of the village brought us an indescribable ambience and mood.

The Qing Dynasty (1644-1911) residence (see Fig. 3) in Shexian County, southern Anhui Province, looks particularly captivating in the shade of trees while bathed in moonlight. Light penetrating windows makes it look as if the house had bright eyes, adding some dynamism to the tranquil scene and creating a poetic illusion. The seemingly plain contour of the building is inspiring. The environment and ambience

山墙、凌空耸起的木质吊脚楼、简易朴素的小棚、各具特征的宅邸和错落有致的屋顶。在雄鸡的鸣唱中，小镇迎来新的一天。

当我们来到皖南古镇或贵州石板寨时，往往会不自觉地受到民居美的感染，如饮醇醪，似醉其中，赞叹其美。从美学角度来讲，在现实的审美世界中，民居并不是以类型，而是以个别的形式展现出来的。不同的民居具有不同的形式，给人的感觉也各有异采。

皖南某宅（见图二）完全保持明代民居的风格，它给我们的感觉是肃穆、古朴，在层层下落的屋顶中，我们感觉到节奏的律动，在以横线为主的轮廓线中，我们感觉到恬静。温庭筠《商山早行》中有两句名诗："鸡声茅店月，人迹板桥霜。"把我们带进了一个特定的艺术意境中，被它所感染。我们似乎感受到一种特殊的气氛和情调。而这种感受又好像是"只可意会，不可言传"。

皖南歙县一处清代住宅（见图三）在月光下，树影婆娑，一个个窗子像闪亮的眼睛，流露出绵绵情思，静中有动，动中寓静，把我们的思绪带

图二 皖南某宅
Fig. 2 A mansion in southern Anhui Province.

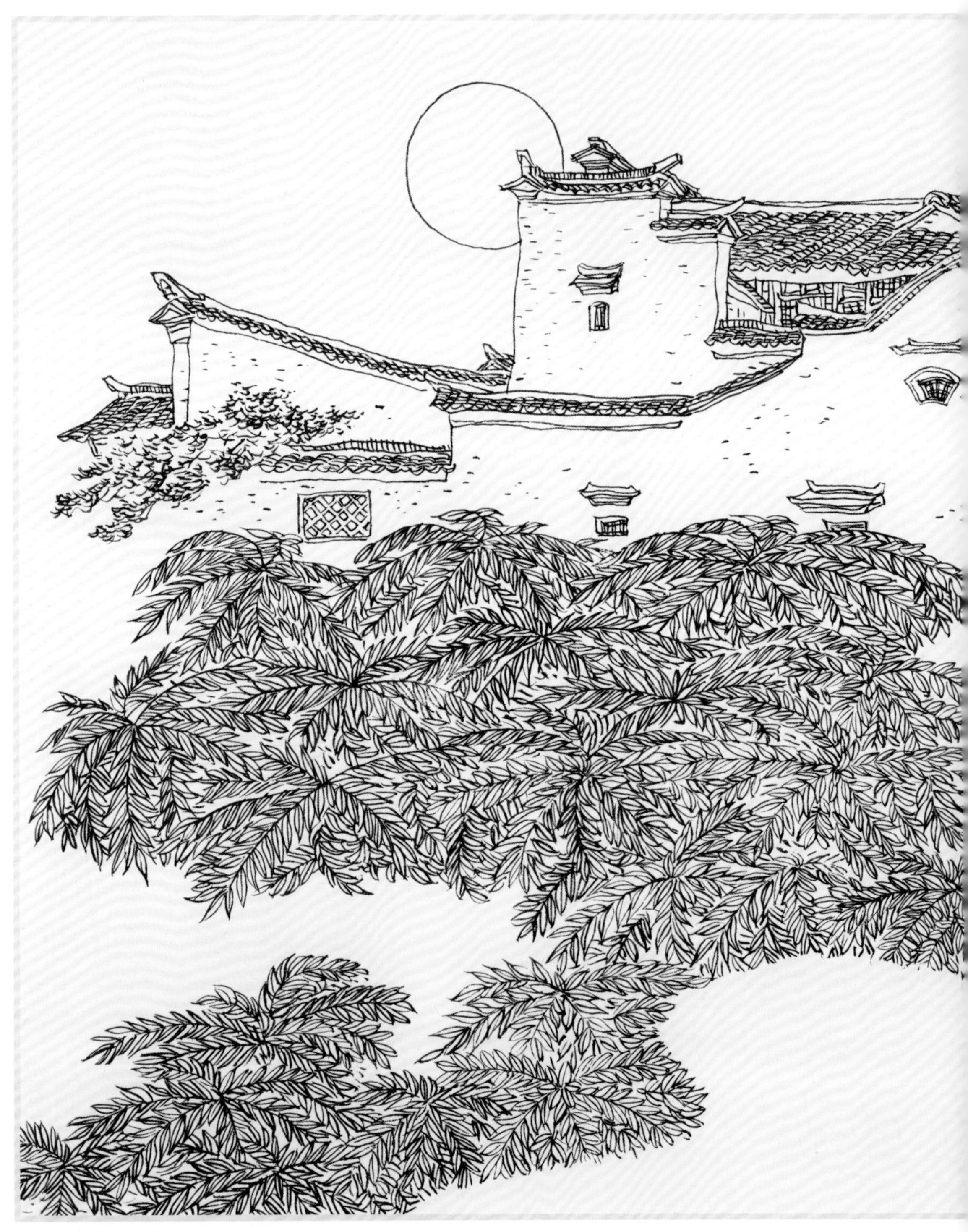

图三 皖南歙县清代住宅
Fig. 3 A Qing Dynasty mansion in Shexian County, southern Anhui Province.

caused an intuitional, blurred feeling. Mei Shengyu, a renowned poet of the Song Dynasty (960-1279), once said, "When an author portrays his inner world in text, readers may understand his meaning but cannot convey it precisely with their own words, even with a general explanation." This is a common phenomenon in aesthetic practice. For instance, when we read a poem, it's often difficult to express its beauty with words. Similarly, it is often difficult to precisely explain the beauty of traditional residences with words.

When we saw residences (see Fig. 4) in western Zhejiang Province, we couldn't help but recall *My Thatched Cottage*, a poem by Song Dynasty poet Lu You: "My thatched cottage is small but exquisite, with a new wooden gate covered with green moss." In a grove flanking a zigzagging mountain path, the melodic singing of nightingales reverberated through the air. The picturesque scenery in early summer made the residence particularly poetic and tranquil.

Traditional residences embody a perfect combination of natural and artistic beauty. Roaming a countryside path or old street in an ancient town seldom fails to deliver a rush of beauty.

入富于诗意的遐想中。建筑物的轮廓似平非平，耐人寻味。环境气氛给我们的感受，是感性的，直觉的，又是朦胧的。正如宋代诗人梅圣俞所说："作者得于心，览者会以意，殆难指陈以言也。虽然，亦可略道其仿佛……"这可以说是审美生活中常见的一种规律性的现象。正如我们读诗一样，我们体验到了诗的美，但往往是"心中所有，口中所无"。要我们用语言来完整准确地描述民居美，亦是同样困难。

当我们看到浙江西部地区的民居（见图四），就会想起宋代诗人陆游《吾庐》在诗中说："吾庐虽小亦佳哉，新作柴门斸绿苔。"蜿蜒起伏的山路边，传来了悦耳的莺声，在明丽秀美的初夏山景中，这幢民居展示出它的清新韵致和盎然画意。

民居之美既是自然美，又是艺术美，更是自然美与艺术美的完美结合。漫步于乡间小路，徘徊于古城老街，我们心中产生了美的感觉和情绪。

图四 浙江西部地区民居
Fig. 4 A traditional residence in western Zhejiang Province.

图五 湘西凤凰县城一角
Fig. 5 A corner of the Fenghuang Ancient Town in western Hunan Province.

At dusk, the ancient town of Fenghuang (see Fig. 5) in western Hunan Province gradually goes quiet as tourists disperse. A stroll along a stone plank road inspires strong nostalgia. Following the zigzagging path to a magnificent gate tower felt like strolling through a painting.

"An old friend prepared chicken and food and invited me to his cottage hall," wrote Tang Dynasty poet Meng Haoran in *Visiting on Old Friend's Cottage*. "The village is surrounded by green wood. Angular green mountains set off the city wall. The windows open to fields and earth; cup in hand, we talk of crops of grain. When the Double Ninth Festival nears, I will come back for more chrysanthemums." The tranquil, poetic rural scenery and the deep respect captured in this poem radiate from the residence (see Fig. 6) on the bank of Taihu Lake.

The harmony, dynamism, tranquility and simplicity of traditional residences makes it easy to forget nature and ourselves in favor of intoxication by indescribable beauty. The greatest admirers of the beauty of traditional residences usually live in modern buildings, while the owners of the residences are often more indifferent. Why? Many aestheticians would point to the impact of "psychological distance."

The more "psychological distance" a person maintains from traditional residences, the more they tend to appreciate their architectural aesthetics. This phenomenon is vividly depicted by Qing Dynasty poet Guo Liufang in *Returning to Changsha by Boat*: "Returning home along the east bank of Rain Lake at dusk, I saw beaded curtains dyed red by the setting sun. I had no idea I lived in such a picturesque place until today I saw my home from the river."

On the highlands near the Jialing River in Sichuan Province are many residences with primitive beauty (see Fig. 7). During autumn, the residences reflect off the water, creating a magnificent, ethereal vista with distant mountains as the background. In Chongqing, nicknamed the "mountain city," we wandered from one building to

日暮人归之后，在湘西凤凰县城的一角（见图五），游人如织的喧闹街道开始冷清起来。踏在石板路上，怀古之情油然而生，那高大巍峨的城楼和曲折延伸的路径，使我们仿佛置身于一幅完美的图画中。

唐代诗人孟浩然的《过故人庄》诗云："故人具鸡黍，邀我至田家。绿树村边合，青山郭外斜。开轩面场圃，把酒话桑麻。待到重阳日，还来就菊花。"这恬静秀美的农村意境和淳朴浑厚的情谊所表现出来的浓郁隽永的诗意，不正体现在这幢太湖边上的民居之中吗？（见图六）

图六 太湖边上的民居
Fig. 6 A traditional residence by Taihu Lake.

在民居的和谐、节奏、静谧、朴实中得到启示，我们忘掉了自然，忘掉了自我的情绪波动和思想起伏，沉浸到美的意境里。大谈民居之美的人，也许住在现代建筑里，而民居的真正主人往往对自己住宅的美不那么敏感。为什么呢？可能是美学家们所说的“心理距离”在此产生效应了吧！

作为观者的自我和民居保持着相当的“心理距离”时，更有可能在心中产生美。这一点在郭六芳的《舟还长沙》诗中就有很准确的说明：“侬家家住雨湖东，十二珠帘夕照红。今日忽从江上望，始知家在画图中。”

在四川嘉陵江边（见图七），高地上蕴藉含蓄的民居，映照在秋江寒水之中，远处的苍山作为背景，使画面的意境愈益宏阔深远。在著名的山城重庆，我们沿石级从房顶走下去，一个个奥妙的空间在前后上下左右变幻着。迷蒙蜿蜒的石阶与大起大落的建筑使人流连忘返。重庆民居（见图八）为什么使我们倍感兴趣呢？

图七 嘉陵江边的民居
Fig. 7 Residences on the bank of the Jialing River.

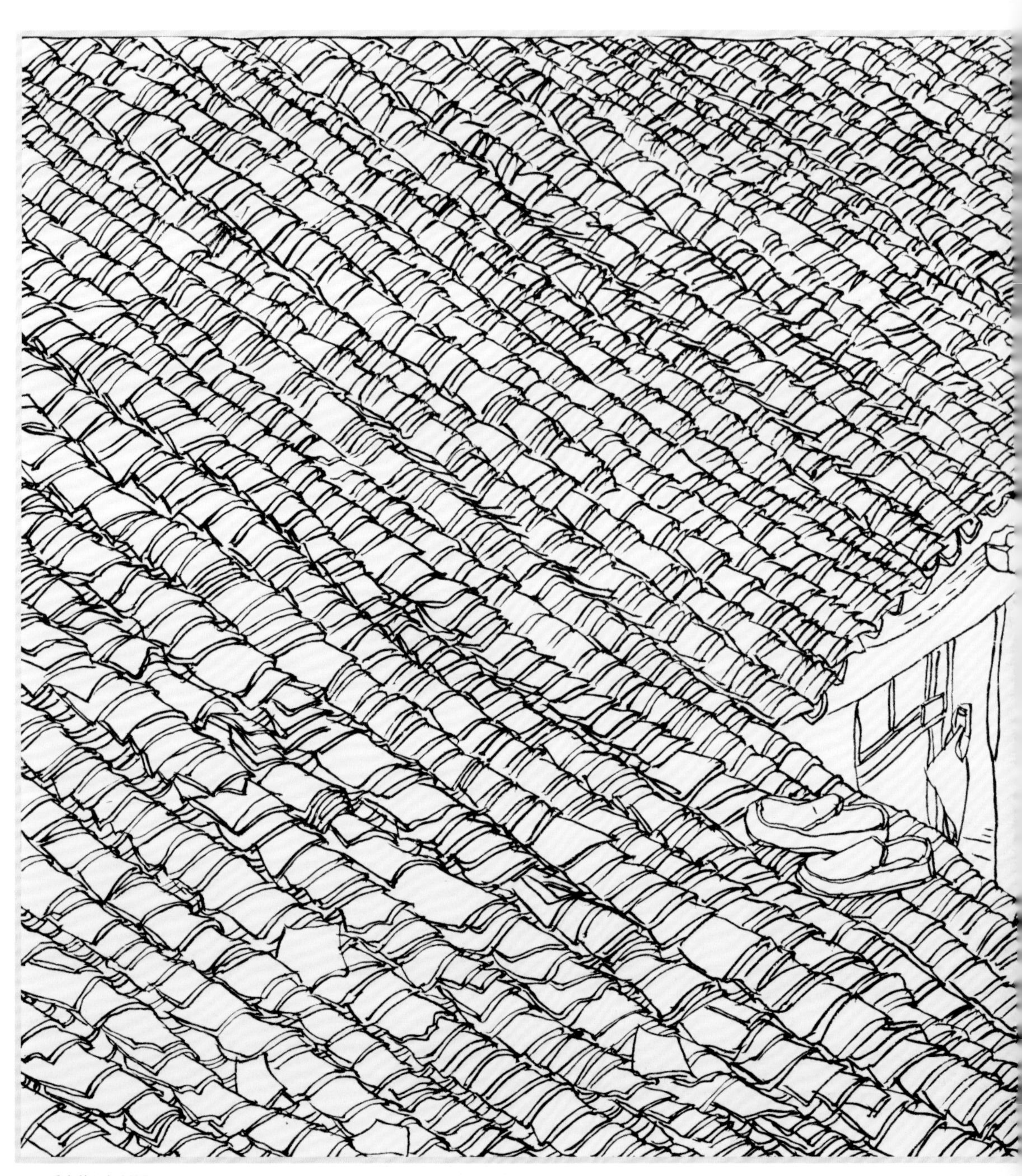

图八 重庆临江门民居
Fig. 8 A traditional residence at Linjiangmen, Chongqing.

another along a snaking stone stairway leading to one amazing space after another. The stone stairway zigzagged into the haze, composing a fascinating picture with surrounding buildings. Why do Chongqing residences (see Fig. 8) deserve particular attention?

Unlike official buildings such as palaces, mansions, temples, and mausoleums, residential buildings were relatively free from regulations and rules in feudal China. For this reason, different regions developed different styles of residences with aesthetic appeal. It isn't hard to see how traditional residences (see Fig. 9) in southern Fujian Province sharply contrast official buildings from the same period thanks to magnificent usage of

图九 闽南民居
Fig. 9 Traditional residences in southern Fujian Province.

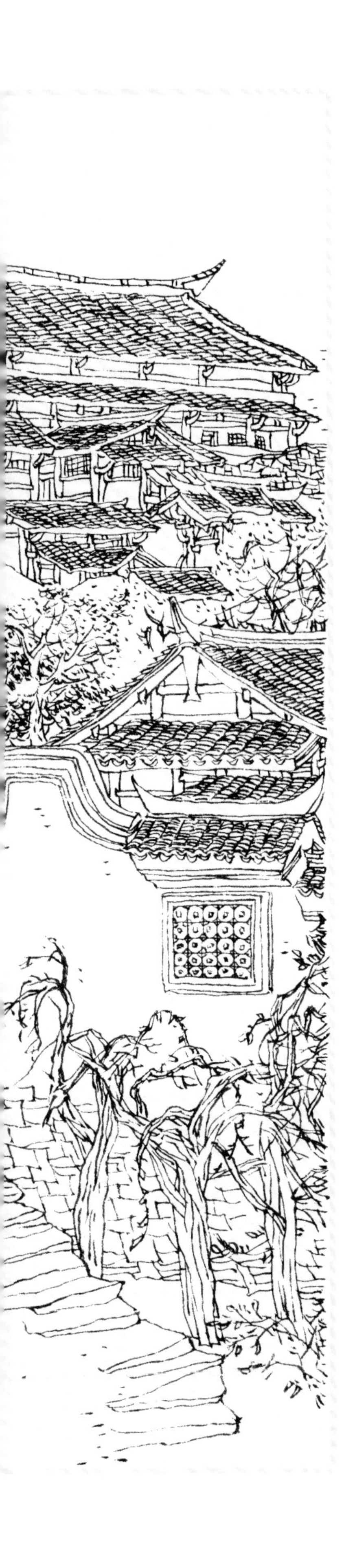

a variety of architectural elements such as saddle roofs, suspension roofs, and double eaves. From the distance, the residential buildings appear like celestial abodes for immortals.

One prominent feature of traditional residences is perfect integration with the surrounding environment. Because builders used indigenous materials, residences maintained harmony with their respective locales. Majestic and resplendent official buildings were like gorgeous Chinese blue-and-green landscape paintings, while traditional residences were more like poetic ink-and-wash landscapes. Homes found harmony with the surrounding environment in terms of the texture of building materials. For instance, local residences on the Loess Plateau were usually yellowish just like the earth. Mountainous areas of eastern Guizhou Province feature grotesque rocks, and local residences (see Fig. 10) built on hillsides feature stone walls and tiles that blend into the rocky world.

这是因为民居和官式建筑如宫殿、府邸、寺观、陵寝不同，受封建统治者规定的“法式”“则例”的限制很少。因此，民居在不同的地点有其不同的形式，很有感染力。从闽南民居（见图九）中，我们不难看出，尽管建筑规模很大，歇山、悬山和重檐都得到了应用，但其突兀峥嵘，意境宏大开阔，完全不同于官式建筑。那石级重楼，远远望去，宛如挂在天幕之下的蓬山瑶台。

美妙的建筑与美妙的环境融为一体是民居的一个特点。由于就地取材，所以建筑的色彩和周围的环境十分协调。官式建筑金碧辉煌，鲜艳夺目，如同国画中的“金碧山水”“青绿山水”。而民居则如同国画中的“水墨山水”，充满诗意，耐人回味。建筑材料的质感，也和周围的环境融和统一。在西北黄土高原，建筑是黄色的。在贵州东部山区，山岩崭露，怪石嶙峋，犬牙交错，建在山坡上错落有致的民居也是石块墙、石板瓦，好一个石头的世界（见图十）！

图十 石板瓦民居
Fig. 10 Traditional residences with stone plank roofs.

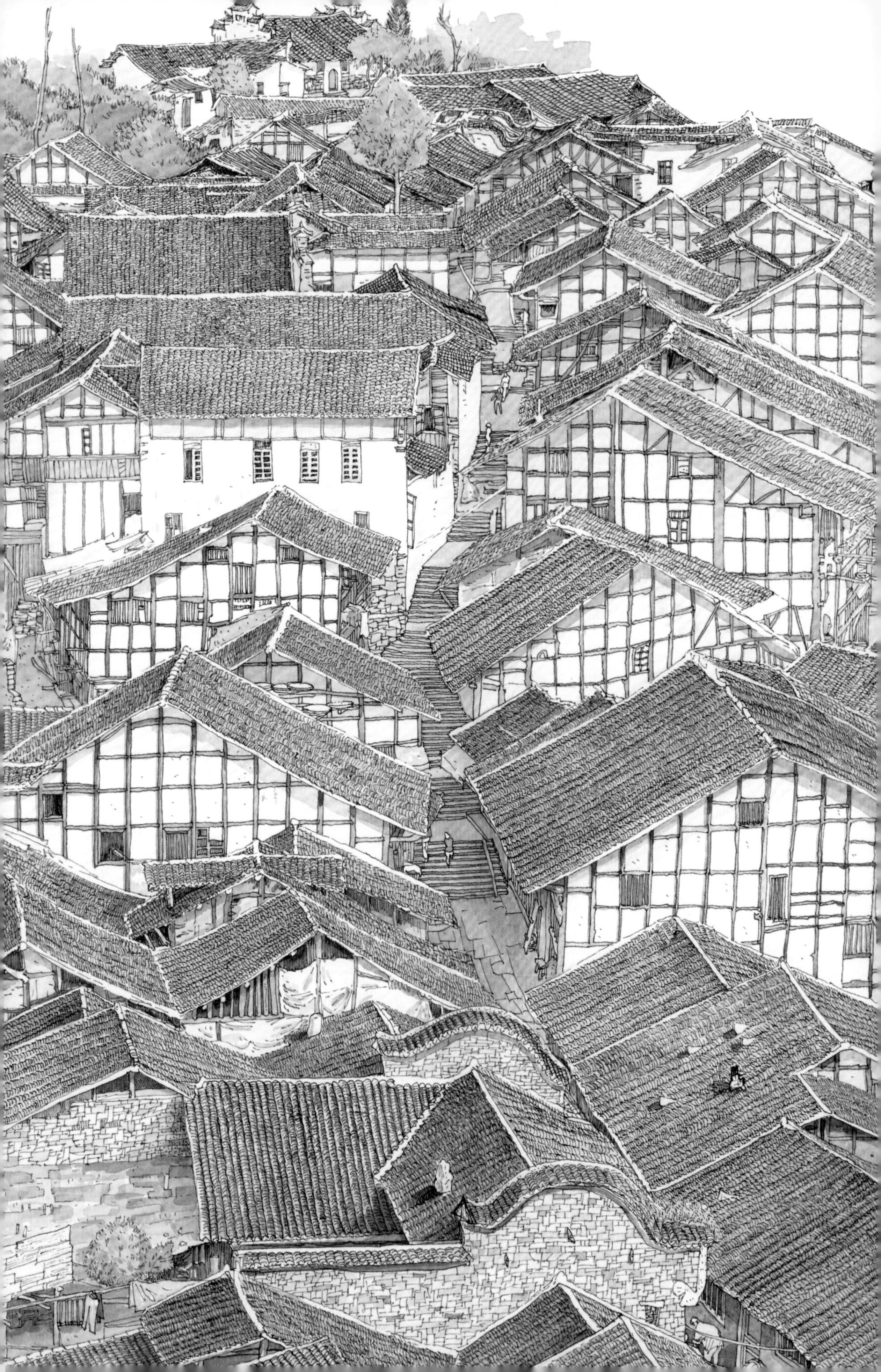

上篇
Chapter I

形式集合

Various Forms of Residences

01 蒙古包 Mongolian Yurts

“敕勒川，阴山下。天似穹庐，笼盖四野。天苍苍，野茫茫，风吹草低见牛羊。”在苍茫原野上，一座座蒙古包点缀着丰草碧波的大草原。蒙古包是蒙古民族固有的居住房屋，即汉代所谓之“穹庐”。其外面用羊毛毡包在简单的木骨架上，平面和屋顶都做成圆形。材料都做成装配式的构件，可以随意安装和拆卸，轻便灵活，方便游牧民族的迁移。蒙古包正中置火炉，烟囱伸出包顶，炉四周为坐卧处。墙壁用木条编成类似篱笆墙的围栏。有的蒙古包是置在预制的圆形火炕上，这样包内就更加温暖了。除蒙古族外，哈萨克等民族为适应游牧生活也使用类似形式的毡包。

蒙古包虽是民居，但不是建筑。不属于建筑的民居还有舟居。船民们以船为家，终日漂游，自得其乐。作为民居，笔者认为这是一种独特的类型。

"The Chile River flows where the Gloom Mountains stand," goes an ancient ballad. "The sky is a dome over the wilds. The sky is blue cast, and the grass extends vast. Cattle and sheep appear out of the grass like a blast." Vast grasslands are often dotted with numerous Mongolian yurts, a traditional Mongolian dwelling. In the Han Dynasty (202 B.C.-220 A.D.), Mongolian yurts were called "Qionglu" (literally "domed abodes"). Such yurts were built by hanging sheepskins on a wooden frame featuring a round roof. All components are prefabricated to be disassembled and reassembled easily. This flexible design facilitates migration of nomadic tribes. A stove is placed in the center of the yurt with a chimney poking through the roof. People sit or sleep around the stove. The wall is like a curved fence woven with wooden strips. Some Mongolian yurts are assembled around a fireplace, making them even warmer. Alongside the Mongolians, other nomadic ethnic groups like Kazakhs also live in yurts.

Mongolian yurts are dwellings that are usually not permanent structures. Boat dwellings are not considered permanent buildings either. Those residing on boats enjoy drifting from one place to another. I believe both represent a special type of residence.

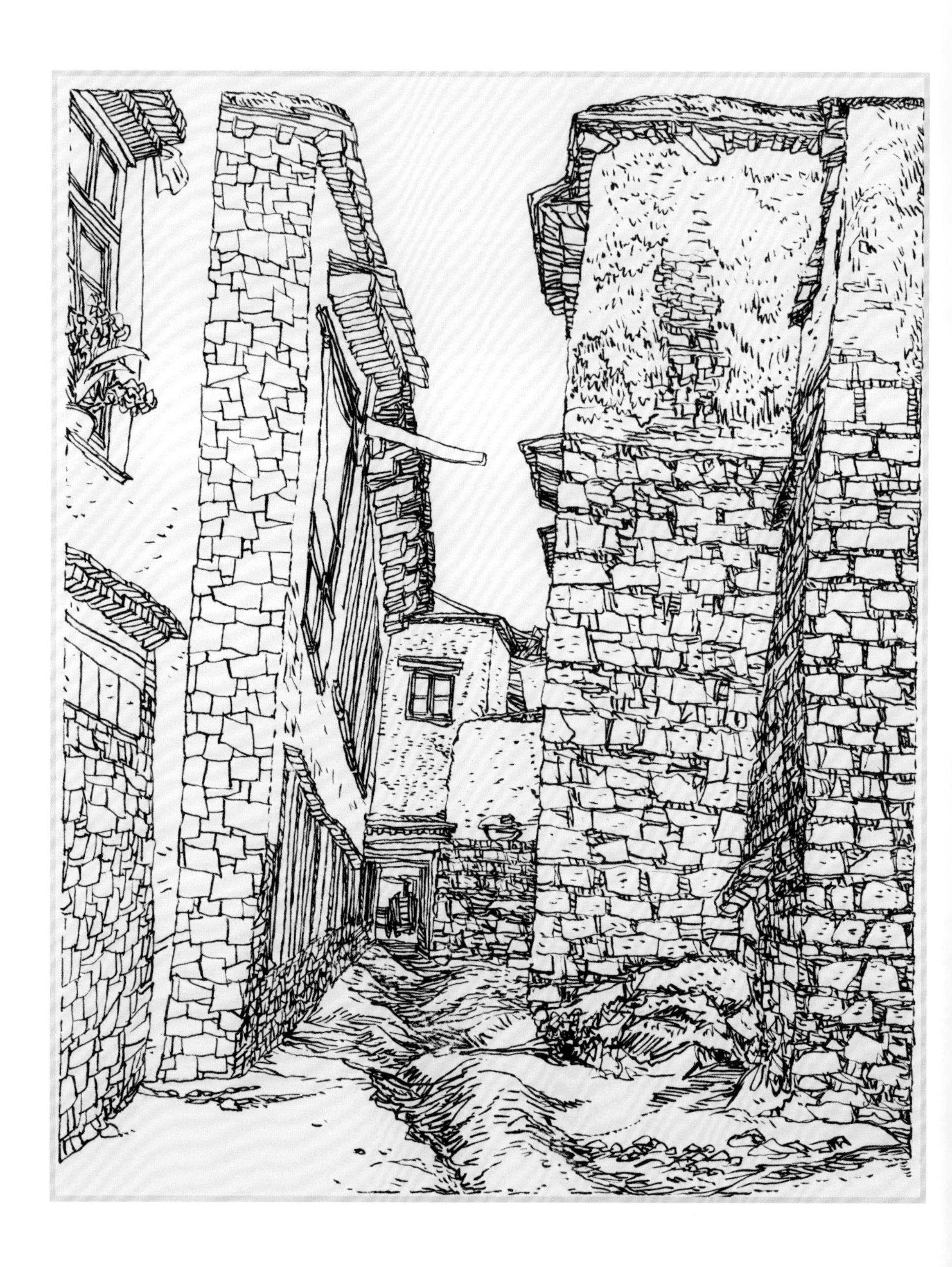

02 藏族碉房
Tibetan Folk Houses

出于防御的考虑，藏族民居以乱石垒砌，高三四层，因外观很像碉堡，故称为“碉房”。这里我们看到的是拉萨某巷民居，森严肃穆，超逸神圣。我们在高耸的建筑下，可以领略到高原的神秘。那深蓝色的天空，那一朵朵仿佛伸手可摘的白云，衬托着藏居的坚实感。藏居的墙体下厚上薄，外形下大上小，建筑平面较为简洁，一般多为方形平面，也有曲尺形的平面。因青藏高原山势起伏，建筑占地过大会增加施工上的困难，故一般建筑平面占地面积较小，转而在空间上发展。

第二十八页图的右上为西藏那曲民居，外形是方形略带曲尺形，中间设一小天井。内部精细隽永，外部风格雄健。高原的日光强烈，民居处于一片银色中，显得格外晶莹耀眼。

第二十八页图的左下为四川马尔康藏居。底层为饲养牲畜及贮藏草料的地方。楼上为起居室、卧室、厨房，也有贮藏之用。顶层有晒台、经堂、晒廊及厕所，经堂的位置最好。最富有创造性的是厕所——挑出墙外，由伸出的搁栅承托，用细树枝编成四周的围墙，粪便直接掉进墙外的粪坑。

For the sake of fortification, Tibetan folk houses are usually three or four-stories tall and built of stone. They resemble blockhouses, hence the name “Diaofang” (literally, “fortified houses”). In a lane of Lhasa, we saw such traditional Tibetan houses looking majestic and ethereal. The towering buildings radiated the mysterious charm of the Qinghai-Tibet Plateau. Under the azure sky dotted with white clouds, the Tibetan residences appeared particularly magnificent and solid. Their walls are thicker in the upper part and thinner in the lower part, and their layouts are usually simple, with most being square-shaped and a few L-shaped. Because of the uneven mountainous topography of the Qinghai-Tibet Plateau, construction becomes exponentially more difficult if more land area is covered. For this reason, Tibetan folk houses usually cover a small area but reach high.

In the upper right of the picture (P28) is a Tibetan folk house in Nagqu, Tibet Autonomous Region. It features a square layout with a small protrusion to form a small “L” shape. In the middle is a small courtyard. The interior decorations are exquisite and have eternal beauty, while the exterior looks magnificent and imposing. Bathed in the sunshine on the plateau, the building glistens with dazzling light.

In the lower left of the picture (P28) is a Tibetan folk house in Barkam, Sichuan Province. The first floor is used for raising cattle and storing food. The second floor consists of a sitting room, a bedroom, a kitchen, and a storage room, and the uppermost floor includes an open-air platform, a Buddhist hall, an open-air corridor, and a toilet, of which the Buddhist hall occupies the most prime location. The most creative feature is the toilet, which projects outwards supported by beams on the wall. Its exterior walls are woven with tree branches, and excrement is meant to fall into a manure pit directly outside the house.

藏族民居在处理住宅的外形上是很成功的。因为简单的方形呈曲尺形平面，很难避免立面的单调。而木质的出挑却以轻巧与灵活和大面积厚实沉重的石墙形成对比，既给人以稳重的感觉，又使外形变化趋于丰富。这种做法兼顾了功能问题和艺术效果，自成格调。

右页图是青海玉树某藏族民居群。吴乔《围炉诗话》说：“情能移境，境亦能移情。”在这种环境里，越发感到大自然的雄奇瑰丽，产生了“精神景观”。精神景观的时间和空间在无形中受到它周围人们的风俗和行为的支配控制，而时空关系的相互渗透和补充又改善、加强了人们的观念，冲击着人们的生活领域，形成民居的一种意境。

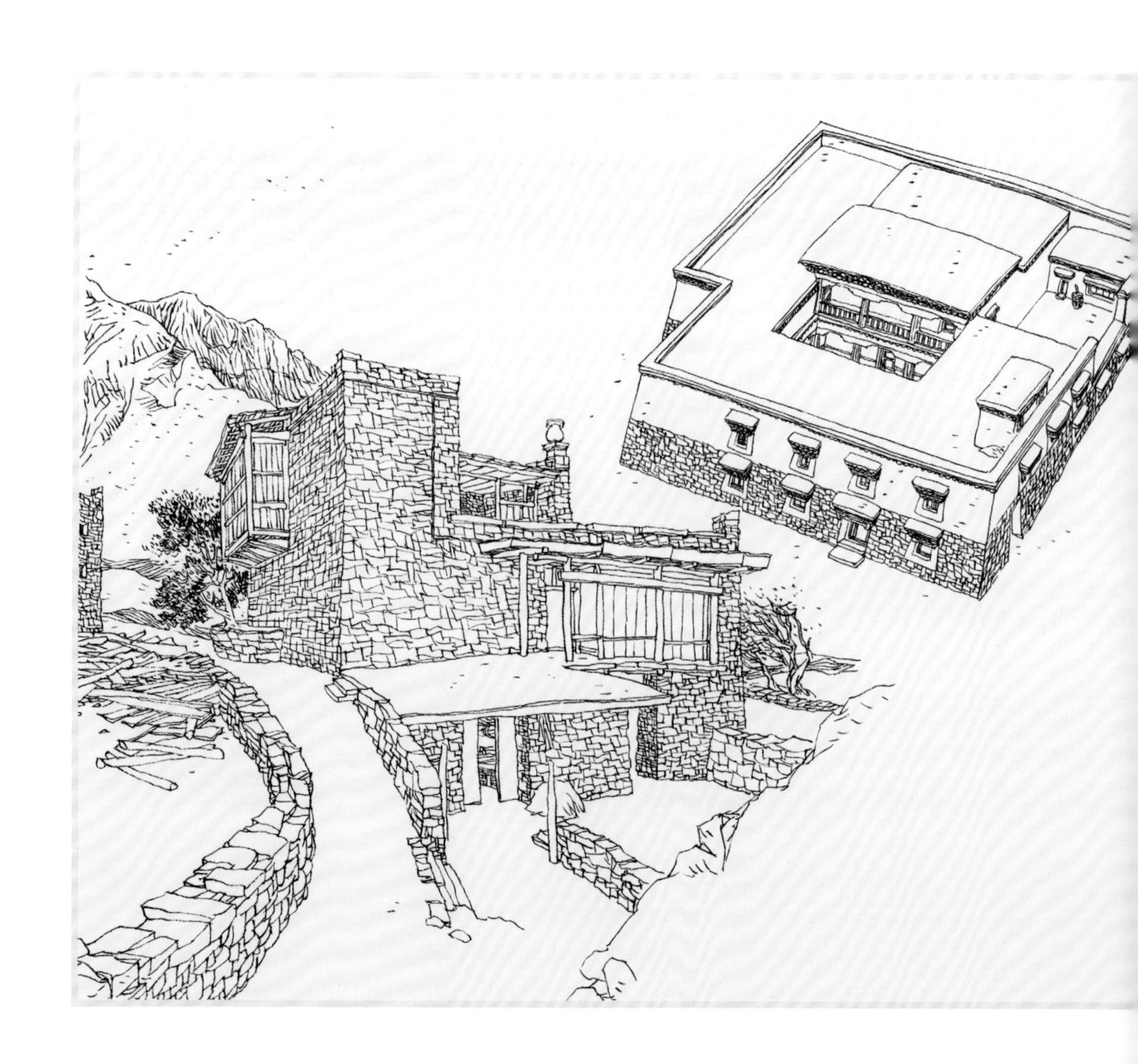

Tibetan folk houses have outstanding exterior design. It is hard for buildings with square or L-shaped layouts to avoid a plain facade. However, through combining light and flexible wooden components and large-size thick stone walls, such houses create a sense of composure while maintaining dynamic appearances. This unique architectural style combines practical functionality and artistic effects.

The picture below shows a Tibetan residential cluster in Yushu, Qinghai Province. Wu Qiao, a scholar from the Qing Dynasty, said in *Weilu Poetry Talks*: "Mood has the power to change the environment, while the environment can also affect the mood." In such an environment, the magnificent beauty of nature captured in a "spiritual sight" can be particularly impressive. The space and time of a "spiritual sight" are unconsciously dominated by the customs and behaviors of local people. The mutual permeation and complementation between space and time improves and enriches the mind and affects lives, creating an ethereal ambience glowing in traditional residences.

民居的每个空间内部都存在诉诸感情和理智的东西。这里，街道是由两侧民居围合的一个带形空间。我们在街道上欣赏和流连，民居成为激起感情波澜的巨石。我们耳濡目染的是藏民热烈奔放的情感，主观与外在的美结合在一起。下图是拉萨八角街一带的街景。我们知道，集镇上的人喜好外部事物，急于感受周围环境，期待与人交往。他们喜欢开敞流动的空间，重视生活形态的时尚。而小村里的人则好沉思，喜内省，无意识中潜伏着对外部影响的抵制，他们偏爱封闭、静止的空间，只在具有历史延续性的环境中生活。这种藏居明显与前面介绍的几种形式不同，其门窗都很大，便于做生意和与人交往，蕴藏着拉萨人直率好客的真情实意和价值观念。

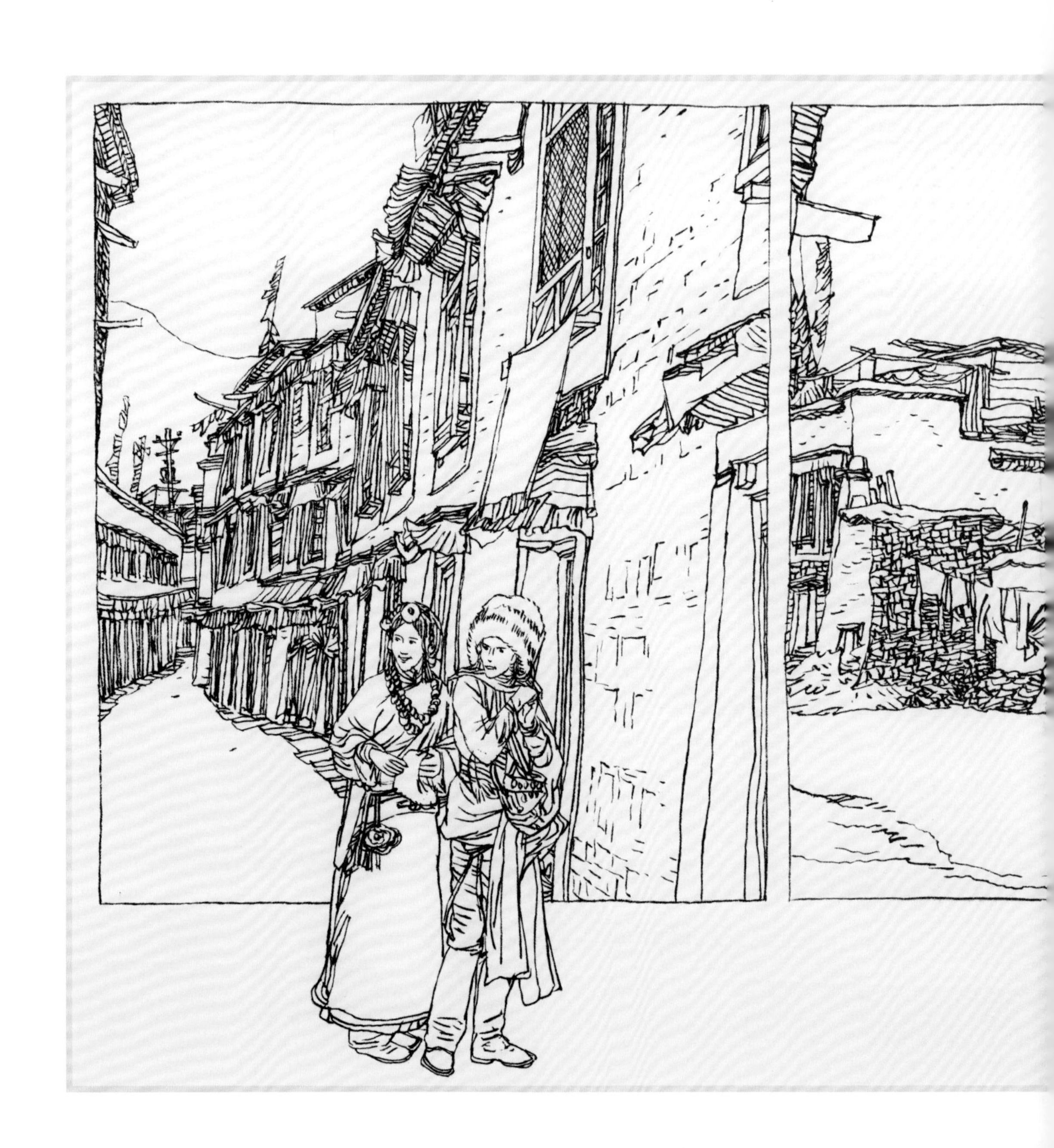

Every internal feature of a residence conveys people's feelings and rationality. A street is a belt-shaped space flanked by residences. As we wander along the street and enjoy its beauty, the residences on both sides evoke distinct emotions. The below picture shows a corner of Barkor Street in Lhasa. In this place integrating internal and external beauty, we were impressed by passionate and hospitable Tibetan residents. Urbanites are often eager to know the outside world, experience the wider environment, and communicate with others. They enjoy flowing and open spaces and value fashionable lifestyles. Rural villagers tend toward more contemplation and introspection and instinctively resist influence from the outside world. They prefer enclosed and quiet spaces and live in a relatively unchanged environment. In contrast to other types of traditional residences, Tibetan folk houses feature bigger doors and windows to facilitate doing business and communicating with neighbors, testifying to Lhasa people's straightforward and hospitable personalities and values.

03 窑洞 Cave Dwellings

右图是陕北延安窑洞，其中夹杂着窑洞式平顶房。右下图是河南巩义市窑洞，从门券形式上可以看出与陕北窑洞的不同。开启的小门叫风门，用竹篾做成，内外糊纸，开启方便且透光，是北方冬季常用的一种门。窑洞是自然图景和生活图景的有机结合，渗透着人们对黄土地的热爱和眷恋之情。

The right picture shows caving dwellings in Yan'an, northern Shaanxi Province, alongside some cave-like bungalows with flat roofs. The lower right picture shows a cave dwelling in Gongyi City, Henan Province. Its arched gate is different from that of the caving dwellings in northern Shaanxi. The small open door is called "Wind Door," woven with bamboo strips and pasted with paper on both sides. It is easy to open and allows sunlight to penetrate. This kind of door is common in traditional residences in northern China and most often used during winter. Cave dwellings represent a perfect combination of nature and life and convey people's affection for the loess land.

现在，我们要沿着洛河西行，去体验广袤的黄土高原。站在高坡上，天际边红日升起，露水打湿了鞋子。阡陌纵横的沟壑把高原分割成无数个土丘，牧童赶着羊群从沟壑中走上来，炊烟在带有寒意的微风中徐徐飘散。沿沟壑而下，啊！窑洞垒垒如蜂房，层叠不穷。

左图是甘肃合水某村。我们看到有的窑洞设在四方形的凹地里，形成“窑洞四合院”。窑洞防火、防噪声，冬暖夏凉，节省土地，经济省工，确是因地制宜的完美建筑形式。“我抬头，向青天，搜寻远去的从前，白云悠悠尽情地游，什么都没改变”。是啊，大地留下我的梦，信天游带走我的情。歌声使我们恍然悟出这里给人以亲切感和奇异吸引力的原因了。

Heading west along the Luohe River, we embarked on a journey across the vast Loess Plateau. As the morning sun rose above the horizon, we stepped onto the highland, and dew wet our shoes. Ditches and valleys divided the plateau into numerous earthy hills. A shepherd watched his flock in a valley as smoke billowed from the kitchens in the nearby village before disappearing into the chilly morning breeze. As we dipped into the valley, a vast expanse of cave dwellings appeared like a gigantic beehive on the hillside.

The left picture shows a village in Heshui County, Gansu Province. Some cave dwellings were built in a sunken square lot to form a *siheyuan* (quadrangle residence). Cave dwellings are resistant to fire and noise. They are warm in winter and cool in summer. Moreover, they save land and are easy to construct, making them an ideal abode in favor of the local conditions. “I look up at the blue sky in hopes of browsing the long gone past,” goes a folk song. “White clouds float carelessly like nothing has changed.” The land carries dreams, and folk songs express emotions. The folk song helped us understand why the land feels so cordial and tempting.

04 朝鲜族民居
Traditional Korean Houses

在突兀秀拔的长白山下，居住着延边朝鲜族同胞。他们依山傍水开发了大量水田，世代在这里生息、繁衍。朝鲜族民居保持了我国唐代以前民居的风格，与日本民居形式相近。屋顶常为庑殿顶。无窗，门的格扇做成落地式，为推拉门。房前均有廊，进屋时把鞋靴脱在廊上，赤脚进屋。居室白天作起居室，夜间即作卧室。室内处处可以席地坐卧，这是唐代以前的生活方式。在东北寒冷地区，一般住宅以厚砖墙、土坯墙防寒冷。而朝鲜民居则用薄墙、大面积火炕来御寒，很有特色。到了夏季，更显其适用性。上图是吉林安图偏廊式住宅。朝鲜族民居绝大多数没有院落和围墙，人际关系和睦，亲如一家。

In Yanbian Prefecture at the foot of the magnificent and picturesque Changbai Mountains live many people from the Korean ethnic group. Generations have subsisted on farming paddy fields. Traditional Korean houses retain the style of residential buildings from the Tang Dynasty and resemble Japanese residences. Typically, a traditional Korean house features a hip roof without windows and sliding doors extending from floor to eave. There is a front corridor outside the house, and dwellers take off their shoes and put them on the floor of the corridor before entering bare-footed. The house serves as a sitting room in the daytime, and the dwellers sleep on the floor at night. The lifestyle is identical to the customs of the Tang Dynasty. Considering the cold winters in northeastern China, local residences usually have thick brick or clay walls to preserve warmth. However, traditional Korean houses are exceptions. They have thin walls but use large heated brick beds to overcome the cold of winter. This design also makes them cooler in summer. The above picture shows a residence with a corridor in Antu County, Jilin Province. The overwhelming majority of traditional Korean houses have no compounds or walls, enabling neighbors to live in harmony like one big family.

东北民居
Traditional Residences in Northeastern China

东北天气寒冷，所以住宅多喜向阳。正房前常用大院庭以便多纳阳光。下图所示的这种住宅在东北三省均可见到。像东北民歌二人转那样，这种民居也具有强烈的地方特色，不同于关内的住宅。冬天，天空是莹白的，无边的雪地竟也变得有些浅蓝色了，遥远的苍山下是一片白屋，走进房屋，冷峻孤独之感顿时消失，熊熊的火炉，暖暖的热炕，温煦舒适，深情绵邈。

Due to the chilly winters in northeastern China, local traditional houses usually have a spacious courtyard in front of the main hall to allow more sunlight to penetrate. The residence depicted below is common in all three provinces of northeastern China. Just like *errenzhuan* (a song-and-dance duet popular in the northeast of China), this kind of traditional residence has strong regional flavor in sharp contrast to folk residences in regions south of the Shanhai Pass. Against a clear sky, the boundless snow-carpeted land looks a bit bluish in winter. Rows of snow-covered houses nestle at the foot of a distant mountain. Entering a house, however, makes the cold disappear immediately, thanks to the comfort of a burning stove and heated brick bed.

上图是吉林北部民居。东北民居常用平屋顶，在檩上置椽铺草巴或秫秸，上面铺碱土、灰土等。这栋民居是在房顶加砌三面女儿墙，前面留一部分小的斜坡屋顶，如同虎头向前伸张，故当地人称此种住宅为“虎头房”。虎头房是富裕人家为适应风大的自然环境并适当美化房屋而建造的，构思新巧，颇有特色，富有生机的造型充满了浓郁的生活气息。女儿墙面有各式透珑花格，精致的花纹，颇悦人目。

The picture above shows a traditional residence in northern Jilin Province. Typically, traditional houses in northeastern China feature flat roofs made of straw covered with alkali clay and lime. The top of the house features parapets in three directions and a single sloped roof on the front shaped like a forward-facing tiger's head. Thus, locals call this type of residences "tiger's head houses." Such houses, which are typically inhabited by wealthy families, are designed to suit the local natural condition of frequent strong winds. The "tiger's head" design is beautiful and creative, adding vitality to the building with a heavy breath of life. The parapets on the roof feature exquisite and attractive engraved patterns in various styles.

06 北京四合院 *Siheyuan* in Beijing

住宅庭院的大小与气候冷暖有关，越是寒冷的地方，院子就越大。四合院四方四正，里面暗含一个“井”字格局。从奴隶制社会的“井田制”到以后发展起来的明堂、宫室、宗庙建筑，中国传统建筑始终力图使建筑艺术具有鲜明的社会性、政治性和伦理性。“井”字分割产生一个中点，中是对称、是稳定、是端庄、是严肃，很容易衍生出许多象征内容。先天八卦北为坤卦（☷），坤为地；南为乾卦（☰），乾为天。南北朝向即“天地定位”“乾坤之事”。顺应天道，自然会大福大吉。在四合院中，北屋是最适合人居住的，但会客、祭祀的厅堂都设在北屋。东西厢房和倒座、后堂才真正用于居住。这说明中国人把理性放在实用功能之前。越是格局讲究的民居，越能体现出这一点。大门、影壁、垂花门、游廊都是为增加气派而设置的。进入四合院，空间序列井井有条，建筑尺寸适度合理。

Typically, the size of a residential courtyard correlates to climate. The colder a place, the bigger the courtyards. The quadrangle layout of *siheyuan* resembles Chinese character “Jing” (meaning “water well”). From the well-field system in Chinese slave society to later structures like halls, palaces, and temples, traditional Chinese architecture has always boasted distinctive social, political and ethical functions. Around the center of the “Jing” shaped layout are symmetrically distributed buildings. This layout represents stability, dignity, and solemnity and encapsulates many symbolic meanings. The north of the Eight Diagrams is called “Kun” (meaning “earth”), and the south is called “Qian” (meaning “heaven”). They together embody the philosophy of “harmony between earth and heaven.” Mankind will be blessed as long as they adapt to nature. In a *siheyuan* compound, the north hall is the best place for dwelling. However, it is typically used as a sitting room or a place to worship ancestors. The rooms in the east, west, and south as well as those in the rear yard usually serve as bedrooms. All this is evidence that Chinese people place rationality above functionality, which is further embodied in exquisitely designed houses. The front gate, screen wall, festoon gate, and wandering corridor add the grandeur of the residence. The spaces inside a *siheyuan* are neatly arranged, and all buildings meet the principle of proportionality.

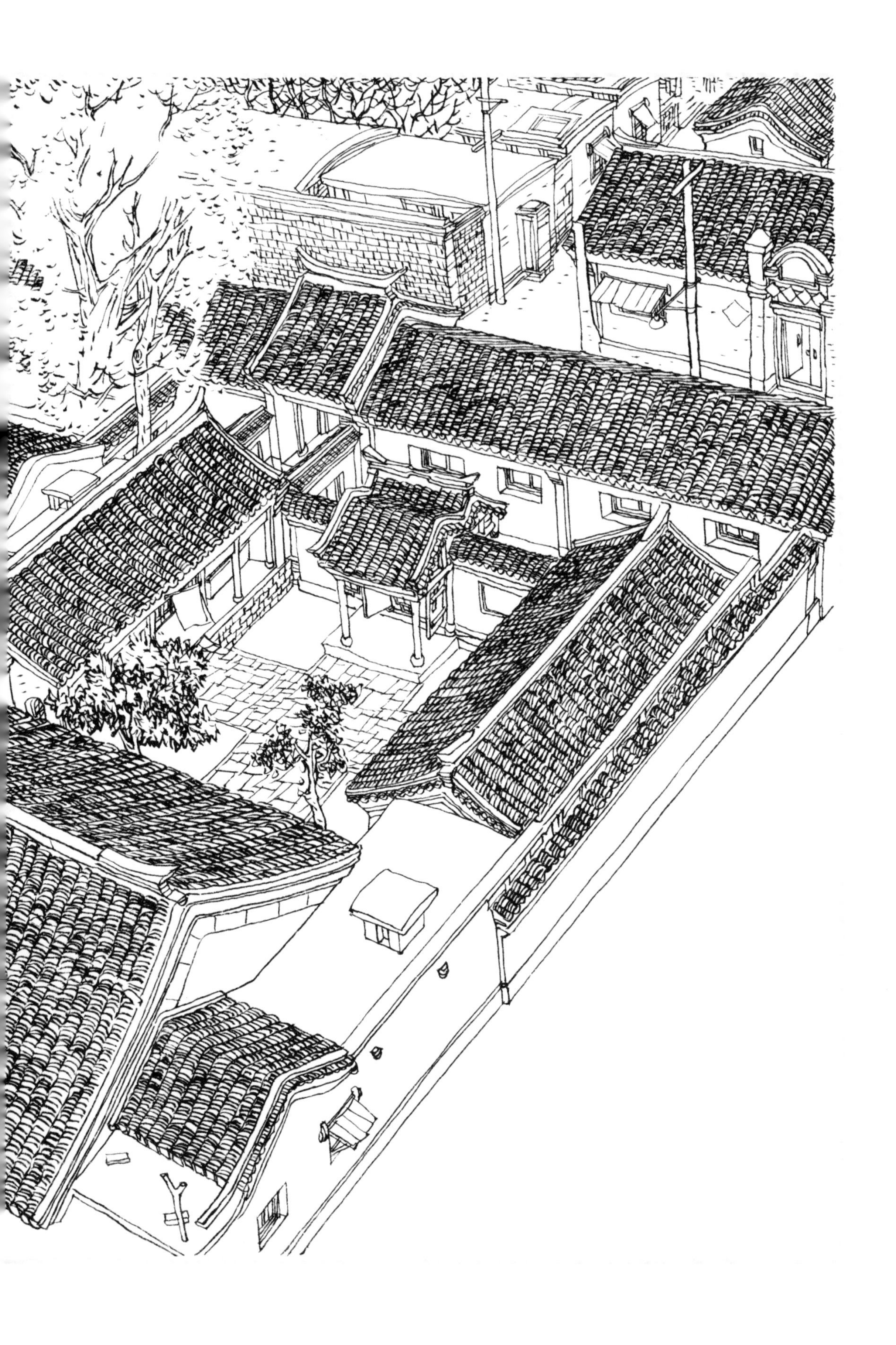

四合院内院由二门起到正房止，常有游廊围绕。游廊不仅遮雨，而且使院庭产生了回合，有波澜跌宕的意趣。大门一般设在左前端，进门即是照壁，院内幽深寂静，外人看不到，于是天热时庭院可以作为一个很好的户外起居室。院庭内往往铺砖地，摆一些盆花和盆景，院庭兼作小园。在左侧两幅图片中，上图为北京东城区某胡同；下图是四合院内景。从中可以看出稳定而端庄的轮廓，均衡而严肃的格调。

The rear courtyard is usually encircled by corridors wandering from the second gate to the main room. The corridors not only shelter people from rain, but also add a dynamic beauty to the courtyard. Typically, the front gate is in the left front of the courtyard, facing a screen wall inside. The screen wall blocks the view of pedestrians, making them unable to see the inside of the courtyard. The tranquil courtyard serves an ideal outdoor space to escape summer heat. The floor of the courtyard is usually paved with brick tiles and decorated with potted plants, making it look like a garden. The above picture on the left shows a lane in Beijing's Dongcheng District, and the lower picture shows a *siheyuan* featuring composed and elegant corridors and demonstrating a balanced, solemn touch.

江浙民居 Traditional Residences in Jiangsu and Zhejiang

与凝重的北京四合院相对比的是活泼的江浙民居。浙北与苏南位于太湖流域，这里气候温润，无严寒酷暑，唯夏季有一段湿热的梅雨季节。在这种良好的自然条件之下，房屋的朝向多为南或东南。这一地区民居都为木架承重，屋脊高，进深深，防热通风效果好。另外，在平面的处理上尽可能采用置小天井及前后开窗的做法。门窗基本采用低的槛窗及长格扇窗。右图是浙江嘉善某宅，图中有四坡水、封火山墙、悬山、硬山、披屋等多种建筑形式，给人留下深刻印象的是凝练而明确的线条，亲切而和谐的节奏。

Compared to solemn *siheyuan* in Beijing, traditional residences in Jiangsu and Zhejiang provinces are more dynamic. The area around Taihu Lake in northern Zhejiang and southern Jiangsu features a warm, humid climate. The place escapes both scorching summer and chilly winter except for about a month of a rainy season that is also hot. Due to the favorable climatic conditions, most houses in the region face south or southeast and are supported by wooden frameworks. They feature high roofs and large depth. This design facilitates ventilation, making the houses cool in summer. They usually have a small courtyard with low sill wall windows or long lattice windows on both sides. The right picture shows a traditional residence in Jiashan County, Zhejiang Province. It combines diverse architectural elements including four-side sloped roofs, overhanging gable roofs, gable roofs, and penthouses. The compact and clear outlines emit an amiable and harmonious aura.

纵观全国民居，依笔者之见，苏北、皖北、豫东和鲁南接壤的淮海地区民居是建筑形式和平面布局最单调的地区，基本都是硬山式平房，且草顶居多。除前面设窗外，其余三面均不设窗。建筑群的组合虽有三合或四合院，但更多的是分散式布局。同处华东地区的江浙民居无论是造型还是平面处理，都变化繁多，质量普遍很高。江浙民居以不封闭式为主，平面与立面的处理非常自由灵活。悬山、硬山、歇山、四坡水屋顶皆应用。右图是被称为“东方威尼斯”的苏州的某临河建筑，淡雅的水边景色是那么柔和幽静，又隐含着微微漂浮、缓缓流动的意态。诸多画家、诗人曾来此描绘吟颂这“日出江花红胜火，春来江水绿如蓝”的江南景色。

I believe that folk houses in the Huaihai area covering northern Jiangsu, northern Anhui, eastern Henan, and southern Shandong have more monotonous architectural styles and layouts than traditional residences in other parts of China. Most are bungalows with thatched gable roofs. There are no windows on any of the three walls other than the front. Some are compounds with houses on three or four sides, but most feature randomly distributed houses. However, traditional residences of Jiangsu and Zhejiang in eastern China are diverse and complicated in terms of architectural design and layout, with higher construction quality. Most are non-walled and adopt flexible layouts and facades. They may use a variety of roof types including suspension roof, gable roof, saddle roof, and four-sided slope roof. The right picture shows riverside residences in Suzhou, nicknamed the "Venice of the East." The elegant, tranquil river views create an ethereal and flowing sense that has inspired numerous painters and poets to create works depicting the fantastic spring scenery of southern China. "At sunrise, riverside flowers become redder than fire," goes one famous poem. "In spring, green waves turn as blue as sapphire."

和其他地区一样，经济条件好的人家，住宅在平面上采取对称的布局，四周围高墙封闭，并附以花祖堂，造成曲折变化、主次分明的平面布局。木架结构用正规梁架，厅堂也有用“草架”的。建筑力求采用高档材料，细部装饰华丽，建筑面积也大。江浙民居棱角笔直，严格精确，无笨拙臃肿、敷衍堆砌、形象粗糙之感。精湛的施工技术，使建筑大为生色。左图是浙江绍兴某宅的过河廊，这种过河廊是大户人家连接河两岸住宅的通道。它不仅给房主带来使用上的方便，而且给河道带来了空间上的变化。

Just like in other regions, wealthy families lived in quadrangle residences with a symmetrical layout with gardens and ancestral halls adding dynamics and changes to the architectural layout. Typically, houses have beam-supporting roofs, and some halls have thatched roofs. Houses built with high-end materials are spacious and boast exquisite details and decorations. Traditional residences in Jiangsu and Zhejiang are characterized by straight and strict lines and exquisite designs, making them stand out against cumbersome, perfunctory and rough buildings. Superb construction techniques add luster to the abodes. Pictured is a corridor over a river in the traditional mansion in Shaoxing City, Zhejiang Province. This kind of corridor was used by wealthy households to link houses on opposite sides of a river. Not only does it facilitate easy crossing, but causes spatial changes to the river course.

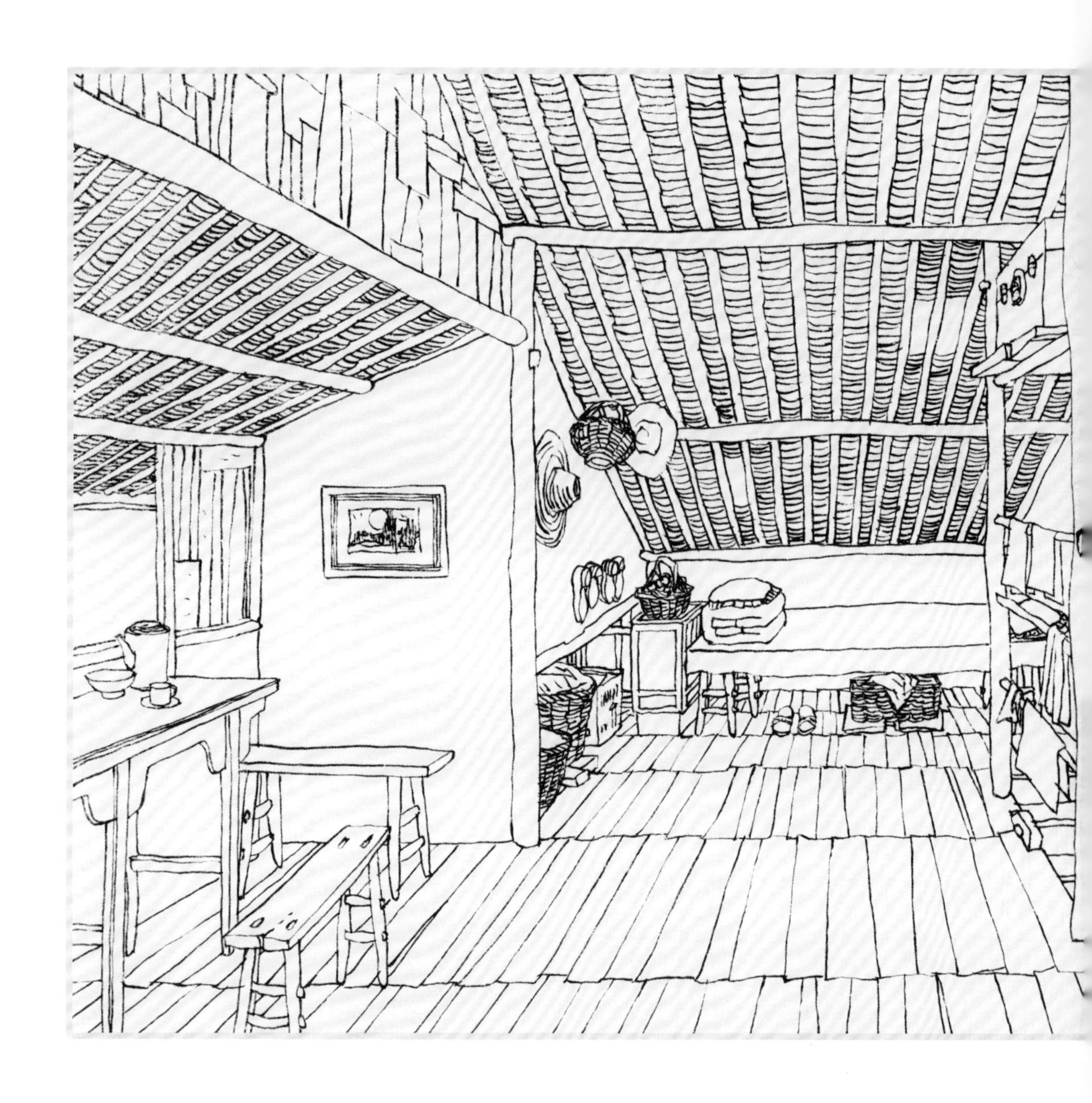

通过视知觉的体验，江浙民居的室内设计往往能激发人的美感。因用途的不同而自然产生的复杂的空间层次，使各房间之间相互联系，浑然一体。其序列不仅合乎逻辑、讲究效能，而且在视觉上惹人注目，功能安排极为合理。别具匠心的空间利用，大大丰富了建筑物的视觉效果。上图左侧为一顶楼内部，仔细观察会发现许多空间利用的方式。上图右侧为一俯视图，从阁楼向下看去，会发现上下楼时必须将楼梯开启，左边出挑的窗台可以睡人，两边的墙壁也被充分利用，挂置衣物。空间利用得如此充分，令人叫绝。

The interior designs of traditional residences in Jiangsu and Zhejiang can be visual feasts. Due to their diverse functions, interior spaces often feature complicated systems connecting different rooms. Arrangements are logical, efficient, and eye-catching, and each part has reasonable function. The innovative utilization of space greatly enriches the visual effect of the buildings. The upper left picture shows the inside layout of an attic. Look carefully to see how rationally the space is used. The upper right picture shows the house viewed from the attic. One needs to pull down a retractable stairway to access the attic. The projecting windowsill on the left can serve as a bed, and the walls on both sides are used to hang clothes and accessories. The smart use of space is impressive.

江浙民居墙身薄，大木结构高瘦，装饰玲珑，木刻砖雕十分精细，屋面轻巧，造成了明秀轻松的外观。白墙黑瓦在丛林溪流的映照下，给人以明快的感觉，素雅清淡，韵味十足。左图左边为江苏吴县扬湾镇景。左图右边为浙江东阳某宅。

Traditional residences in Jiangsu and Zhejiang feature thin walls, slim and tall wooden framework, exquisite decorations, and beautiful wood and brick carvings. The roofs look light, creating a lively and carefree sense. Against lush trees and flowing streams, houses with white walls and black tiled roofs look bright and elegant and blessed with a poetic touch. The left picture shows the ancient town of Yangwan in Wuxian County, Jiangsu Province, and the right shows a traditional mansion in Dongyang City, Zhejiang Province.

08 四川民居
Traditional Sichuan Residences

夕阳西下时的巴蜀之地——四川，深蓝色的天空中，金黄色的星星点缀出各种各样的图案，使这个多山地区充满了神秘感。盆地气候使得这里夏季炎热，冬季少雪，风力不大，雨水较多，于是平房瓦顶、四合头、大出檐成为民居的主要形式，阁楼亦成了贮藏隔热之处。右图是中江地区街景。

Sichuan sunsets are known for golden stars forming various patterns against the dark blue sky, adding a mysterious touch to the mountainous region known as Bashu in ancient times. Located in a basin, Sichuan is hot in summer and has little snow and wind but frequent rainfall in winter. Due to the climate, local folk houses are mostly bungalows with tiled slope roofs, overhanging eaves, and attics used to store sundries and resist heat. The right picture shows a street in Zhongjiang area.

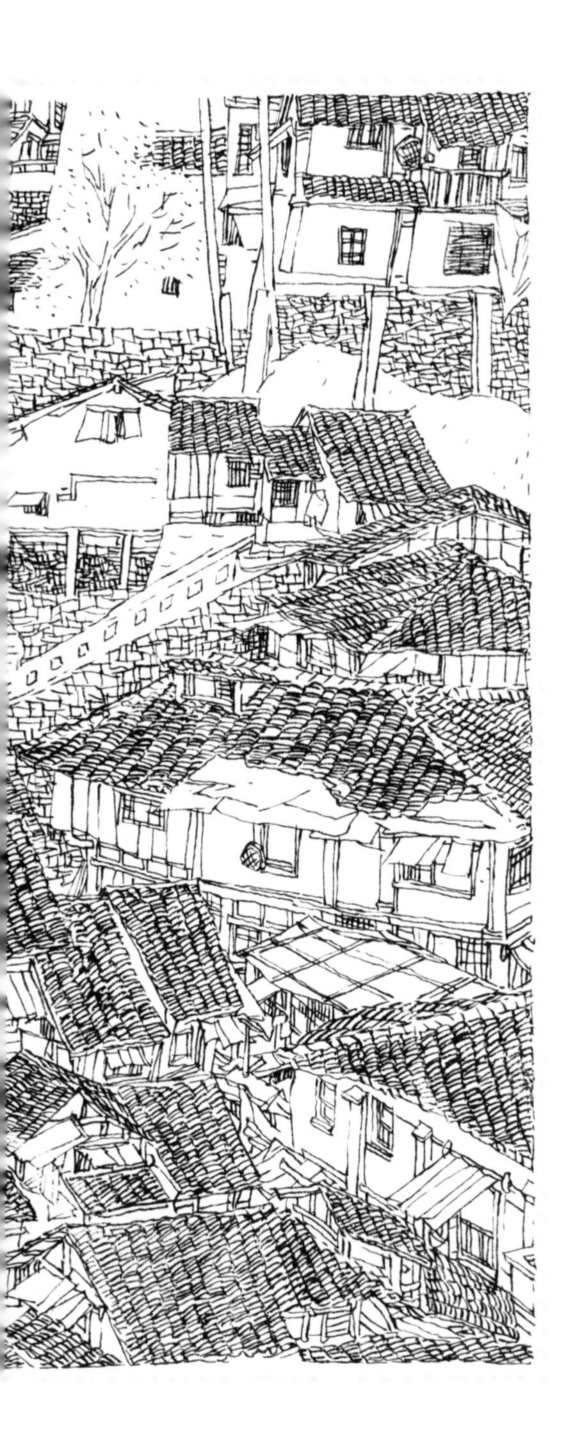

由于多山，山区民居不十分讲究朝向，因地而宜，且天井纵深较浅，以节省用地面积。以图中的北碚某宅为例，四合院屋顶相连，雨天可免受雨淋之苦，夏日也不致使强烈的阳光过多射入室内。而且宅出檐及悬山挑出很大，也可防止夹泥墙或木板墙、桩土墙遭雨水冲刷。

这是重庆临江门民居群，吊脚楼高耸其上，蔚为壮观，穿行于其间的小巷时，就像步入了迷津，有时已经感觉到前边无路可行，但只要侧着身子穿过一条窄缝，便又有另一番天地展现在眼前，真是“山穷水复疑无路，柳暗花明又一村”。

Residences in mountainous areas usually face various directions according to topographical conditions. Moreover, they have smaller courtyards to save space for buildings. For instance, the roofs of the buildings in the Beibei quadrangle residences pictured are interconnected to shelter people from rain and strong sunshine in summer. Meanwhile, its overhanging eaves and gables can protect clay, wooden and rammed walls from the erosion of rainwater.

These dense stilt houses in Linjiangmen, Chongqing, together compose an impressive vista. The surrounding narrow alleys feel labyrinthine. When a dead end seems straight ahead, squeezing through a narrow alley usually brings pedestrians into a brand-new space. Just as a proverb goes, every cloud has a silver lining.

四川民居多为穿斗式屋架。这里的人们在建造民居时善于利用地形，因势修造，不拘成法，常常在同一住宅中，地坪有数个等高线。宅基地的退台有横向有纵向，造成房顶高低的配合。加上一般房檐不高，绿影婆娑，使人感到舒适而明快。重庆及川东山区的民居不注重朝向，依山崖而建，吊脚楼伸出很大，有的层层出挑，气魄宏大，雄伟异常。在上面图片中，左图为万县民居，右图为南充民居。

Most traditional residences in Sichuan feature a through-tenon frame. Locals excel at building houses according to the actual terrains instead of being confined to a certain style. A house can have floors of different contours. The horizontal and latitudinal rooftop platforms at different heights set each other off. Under trees, the relatively low houses feel cozy and lively. Traditional residences in Chongqing and eastern Sichuan are built on the hillsides facing different directions. Some stilt houses project outwards, with each floor overhung on the hillsides which can look impressive and magnificent. The above picture on the left shows residences in Wanxian County, and on the right are residences in Nanchong City.

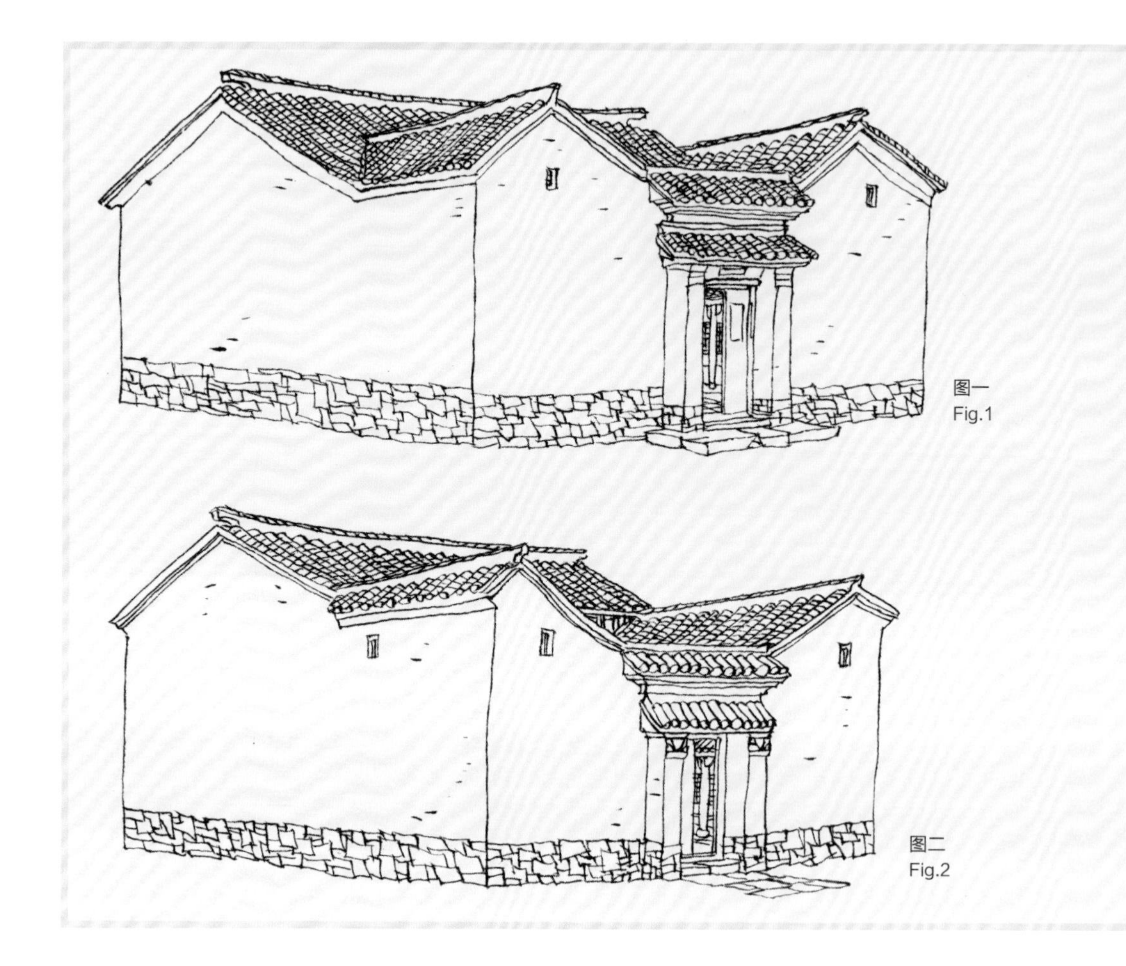

图一
Fig.1

图二
Fig.2

09 云南一颗印民居
Seal-like Compounds in Yunnan

云南地处高原，气候适宜，无严寒酷暑，只是大风较多，故民居都采用厚土墙及筒瓦铺屋顶。一颗印民居因平面方方如印而得名。三间四耳一颗印式是当地最常用的宅制。所谓三间四耳，即正房三间，耳房左右各两间。这种形式的形成除生活生产功能上的需要外，亦与防卫有关。一颗印民居均为楼房，牲畜杂物在楼下，人住楼上。正房楼下是堂屋，作为起居待客之处，堂屋左右作卧室，楼上的中明堂作佛堂。较大的住宅采用两三个一颗印式排列起来。较好的住宅，入门常带有倒八尺倒座。这里介绍一颗印民居的几种外观形式：两耳房双坡屋面（图一）、两耳房单坡屋面（图二）、单耳房（图三）、无耳房（图四）。

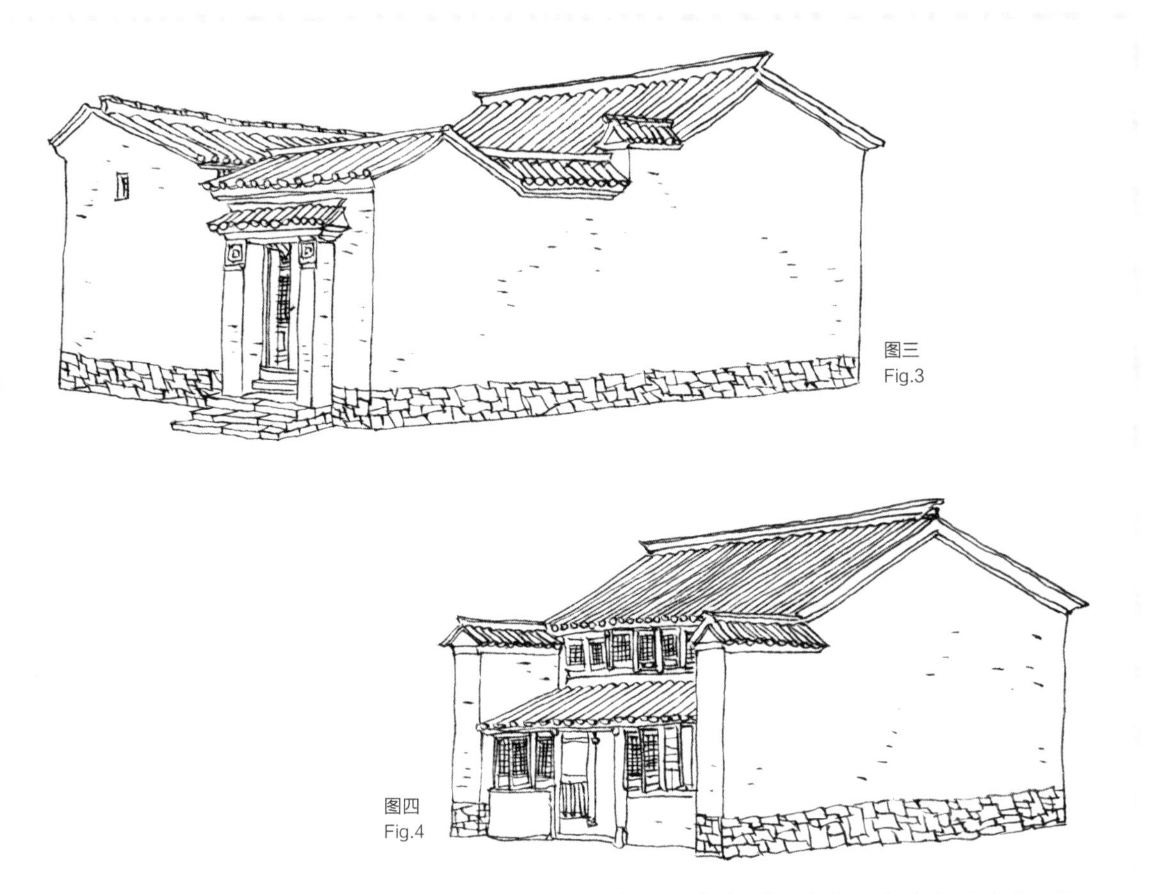

图三
Fig.3

图四
Fig.4

Located on the Yunnan-Guizhou Plateau, Yunnan Province is known for a pleasant climate with mild summers and winters. However, strong winds are frequent in the region. For this reason, local traditional residences feature thick clay walls and tube-shaped roof tiles to resist wind. The square layout inspired the name "seal-like compound." The most typical consists of a three-room main house flanked by a two-room side house on each side. This layout not only meets the needs of daily life and production, but also has a defensive purpose. A seal-like compound is composed of multi-story buildings. The first story is used as a cattle pen and storage, while people sleep above. The first floor of the main hall serves as the sitting room. The middle room on the second floor is used as a Buddhist hall. A large mansion may consist of two or three seal-like compounds. Seal-like compounds of the highest standards often include a north-facing house eight *chi* (about 2.7 meters) in depth. Typical styles of seal-like compounds include those featuring two-room side houses with saddle roofs (Fig.1), two-room side houses with single slope roofs (Fig.2), single-room side houses (Fig.3), and some without side houses (Fig.4).

10 长方形土楼 Rectangular *Tulou*

长方形土楼主要位于福建省西南角与广东毗邻的永定县山区内，在龙岩县也有。土楼的形成与防御外敌有着密切的联系。客家人是从西晋（265—317）时期起由黄河中游一带的汉族人逐步南迁到现在偏僻山区的人群，聚族而居，集体防御，于是由单家小屋建成连居大屋，进而建成多层高楼。长方形土楼有“五凤楼”和普通长方式土楼两种。“五凤楼”式住宅一般由“三堂两落”组成，“三堂”是位于中部南北中轴线上的下堂、中堂和主楼，“两落”是分别位于两侧的纵长方形建筑。当地又称作“横屋”。长方形土楼采取左右对称的布局方法和前低后高的外观，而且大都选在前低后高的地势。它的外观，正面采取对称方式，而侧面则是高低错落的不对称形状。小型的“五凤楼”有的不带横屋，但土墙承重三四层高的主楼是必不可少的。下图是富岭乡的“五凤楼”式住宅“大夫第”，屋顶十分成功地采取了歇山与悬山的巧妙配合，院落重叠，屋宇参差，配以巨大出檐的九脊顶，无论从哪一个角度来观察都显出古朴、庄重、壮观的艺术风格。整个建筑群布局规整、条理井然、主次分明、和谐统一。

Rectangular *tulou* (literally "earth tower") are mainly found in Yongding and Longyan counties in the southwestern tip of Fujian Province along the border with Guangdong Province. The origin of *tulou* was closely associated with defensive needs. In the Western Jin Dynasty (265-317), a group of Han people migrated from the middle reaches of the Yellow River to the remote mountainous region in Fujian. They were later called the Hakkas. Hakkas of the same clan all lived together. For defensive needs, they connected their houses together to form multilayer buildings. Rectangular *tulou* feature two types of layouts: "five-phoenix" style and ordinary rectangular style. Typically, a "five-phoenix" *tulou* consists of "three halls and two belts." The "three halls" are the lower hall, middle hall, and main building on the north-south axis. The "two belts" are rectangular buildings on both sides, which are popularly called "lateral houses" by locals. A rectangular *tulou* usually adopts a symmetrical layout on terrain that is lower in the front and higher in the rear. The front facade is also symmetrical, but its lateral sides are asymmetrical. Some smaller *tulou* in the "five-phoenix" style lack lateral houses, making a three or four-story main building with clay walls necessary. Pictured below is Dafu Mansion, a "five-phoenix" style residence in Fuling Township. Its roof style skillfully combines hip-and-gable roof and suspension roof. The mansion consists of several compounds and many buildings featuring magnificent nine-ridge roofs. From any angle, it radiates a primitive, solemn and majestic aesthetic aura. The entire architectural complex boasts a neat and orderly layout, with the main hall and supplementary buildings coexisting in harmony.

远处的群山在庄重地沉思，周围一片神秘的宁静。在这宁静中，映入我们眼帘的是长方形土楼正立面上部的木板壁与土墙形成的鲜明质感对比，但非常协调。最简单的长方形土楼为“口”字形。最复杂的外围“口”字形楼都高四层，院内的客堂和附属房屋通常仅高一层，当地人称为“厝”。大门位于中轴线上，在中部附属建筑正对中轴线的地方，设供奉祖先牌位的祖堂，祖堂前为大厅，大厅前为迎宾典礼的地点。永定土楼形式多种多样，构成生动而绝妙的艺术形象，屋顶高低错落的变化，丰富了土墙平整的外形。一座座土楼，犹如一个个神奇的城堡，从外面看时，意境高远，哲理深奥，使人非常想进去一游。从小门进入，里面是另一个世界，充满浓郁的生活气息，有的楼房层层出挑，阳台上晒着衣物，整体上层次明晰而富于节奏、尺度紧凑而不失变化。

Distant mountains seemed to get lost in meditation, and everything became mysteriously quiet. In the serenity, the wooden plaque atop the front facade of the rectangular *tulou* formed a sharp but harmonious contrast with the clay wall. The most complicated outer buildings feature a square layout and have four stories, and the living room and other affiliated rooms are usually single-storied, which locals call "Cuo." The main gate sits on the central axis. In the middle of the public building on the central axis is an ancestral hall used to enshrine tablets of the clan's ancestors. The hall in front of the ancestral hall is used to receive guests and hold ceremonies. Various types of tulou can be found in Yongding District, Longyan City. They are lively and fantastic architectural wonders. The rolling roofs going up and down add dynamics to flat and even clay walls. The *tulou* buildings often look like fascinating castles. Viewed from outside, they seem mysterious and rich in philosophical appeal, begging inner exploration. A small gate opens to a different world full of living atmosphere. Some buildings have overhanging balconies on almost every floor, from which dwellers can dry clothes in the sun. In general, they are clearly distributed but full of dynamics and compact in size.

环形土楼
Circular *Tulou*

来到福建省永定区高头乡“承启楼”这座巨大的环形土楼前时，我们便真正领略了其神圣不可侵犯的威严。它建于清康熙年间（1662—1722），外圈周长 229.34 米，高 12.4 米，全楼总面积为 5376.2 平方米。环形土楼的形成起因与设计原则和上面讲到的长方形土楼几乎没有区别，所不同的是它把外部的长方形土楼改成圆形而已。18 世纪（清乾隆）以后，永定县一些烟商大发其财，加上仕官辈出，便大兴土木，永定土楼出现鼎盛时期，毗邻的大埔（广东）、南靖、平和等县也纷纷效仿。环形土楼在遮挡日光的灼射、强风的袭击方面比长方形土楼更为有利。

We were awed by the sacred stateliness of the Chengqi Building, a gigantic circular *tulou* in Gaotou Township, Yongding District, Fujian Province, at first sight. Built during the reign of Emperor Qianlong (1662-1722) in the Qing Dynasty, it measures 229.34 meters in outer circumference and 12.4 meters in height, with a total floor area of 5,376.2 square meters. The origin and design principle of circular *tulou* is identical to its rectangular counterpart. The only difference is the shape. After the 18th century, some tobacco merchants in Yongding amassed considerable wealth, and several talented locals entered officialdom. They built massive houses in their hometown. Yongding *tulou* emerged in this period, and the architectural style was modeled after neighboring counties such as Dapu (in Guangdong Province), Nanjing, and Pinghe. Circular *tulou* are more effective at resisting sunshine and strong winds than rectangular *tulou*.

在内部布局上，环形土楼被分为四等份。定下一条中轴线设大门、祖堂等，在与中轴线垂直的另一条线上设旁门、楼梯、水井等。住宅高度达到四五层。营造者因地取材，用厚度一米以上的夯土筑墙。环形土楼的外墙下部的几层不开窗，有一种稳定感。外墙的淡黄色和上部黑色的瓦顶与苍山清流相掩映，温和、愉快，美丽如画，如同观赏奇异的“仙山楼阁“，使人不得不赞叹我国劳动人民的丰富想象力和创造力。

In terms of internal layout, circular *tulou* consist of four central round structures. The main gate and ancestral hall are located on the central axis, while side gates, stairways, and water wells are positioned along a line perpendicular to the central axis. The buildings have four or five stories. Builders used local materials to make the rammed earth walls as thick as onw meter. The lower part of the exterior wall of a circular *tulou* lacks windows, adding a sense of stability to the entire building. The yellowish exterior wall and the black tiled roof are in harmony with surrounding mountains and rivers, creating a pleasant and beautiful vista like an immortal abode. Upon seeing them, we couldn't help but applaud the imagination and creativity of Chinese laborers.

福建民居
Traditional Residences in Fujian

福建属东南丘陵地带，境内崇山深谷，树木苍郁，气候温暖湿润。山间盛产木材，松杉樟柏等皆有出产，给当地建筑的建造带来了有利条件。福建民居大量使用悬山的人字屋顶，挑出深远的悬山，看过去轻快灵活。在平面呈 90 度角的正房与耳房相连接时，屋顶的处理很有特色，很多是用悬山叠落连接（即老鹰接连）的方式。这是宁德地区某宅，其外观明丽灵活。福建民居的封火墙变化颇多，尤其是悬山顶与封火墙等墙面的配合，常有值得赞赏的优美式样。只要坐车从福州到厦门，沿途就可觉察到福建民居规模大、变化多、细部精的特色。福建民居更多保留了宋代曲线屋顶的特点，从房顶上几乎找不到一条直线。从明间开始，次间、梢间屋檐逐一升起，在屋顶坡度的举折上，每步举高是逐渐升起的，形成凹势圆和的造型。右图为泉州某宅，从房顶可以看出丰富的弧线变化。

A province in southeastern China, Fujian features rolling mountains, deep valleys, lush vegetation, and a warm, humid climate. Its mountainous areas abound in timber resources like pine, cedar, camphor, and cypress trees, creating a favorable condition for local construction. Most traditional residences in Fujian adopt chevron-shaped overhanging gable roofs, which look lively and dynamic. Typically, the roof of the main hall vertically interconnects with wing rooms, forming a distinct style. Many houses feature interconnected overhanging gable roofs (called "eagle interconnection"). Pictured on the right is a stunning traditional residence in Ningde. Traditional residences in Fujian have diverse types of gable walls, and their mingling with overhanging gable roofs forms amazing vistas. Numerous traditional residences of Fujian style line the road from Fuzhou to Xiamen. They have magnificent and diverse designs and exquisite details. Traditional residences in Fujian retain the curved roof ridges of Song-Dynasty buildings, but a few have completely straight roof ridges. The eaves of the middle room, side rooms, and farthest rooms gradually rise upwards, forming a curve lower in the middle and higher on both ends. In the right picture is a traditional mansion in Quanzhou City with dynamic curves on the roof.

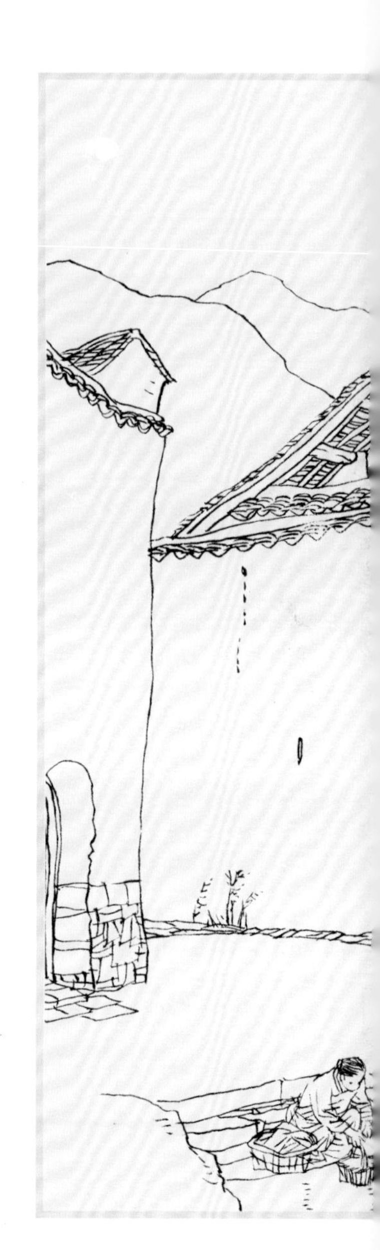

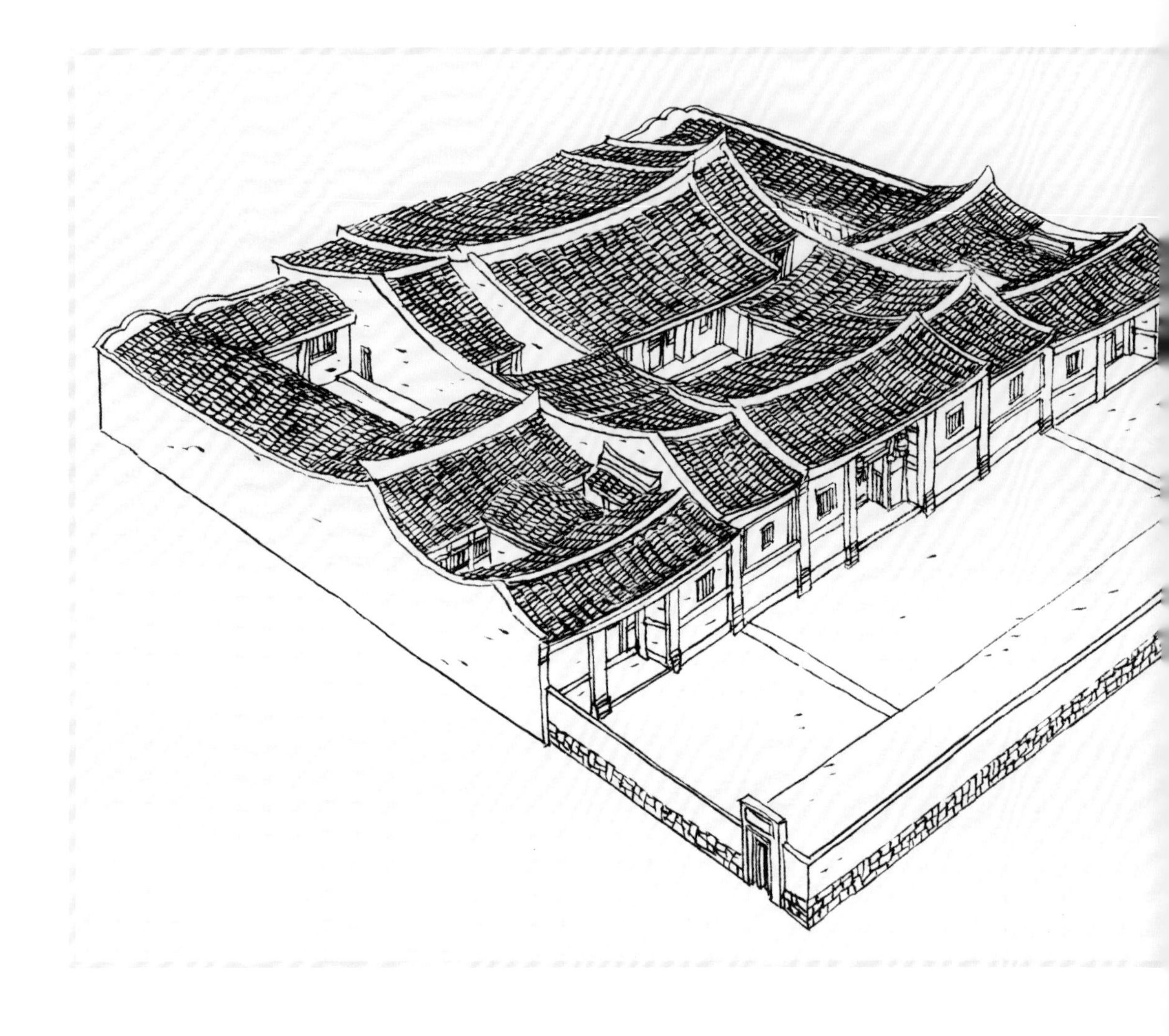

上图右上部分是该建筑的平面图，从中可以看出，主要建筑前面的院子最大，其他的较小，以求主次分明。院子的形状力求产生变化和对比。方形的、纵长的、横长的相互交叉间隔。在可能的情况下，各进院子内建筑物的地平线自前而后逐步地提高，以增加建筑物的庄严感。在同一组建筑群中，由于地形或其他原因而不能用一根中轴线贯通一气时，就分成几段连接起来。

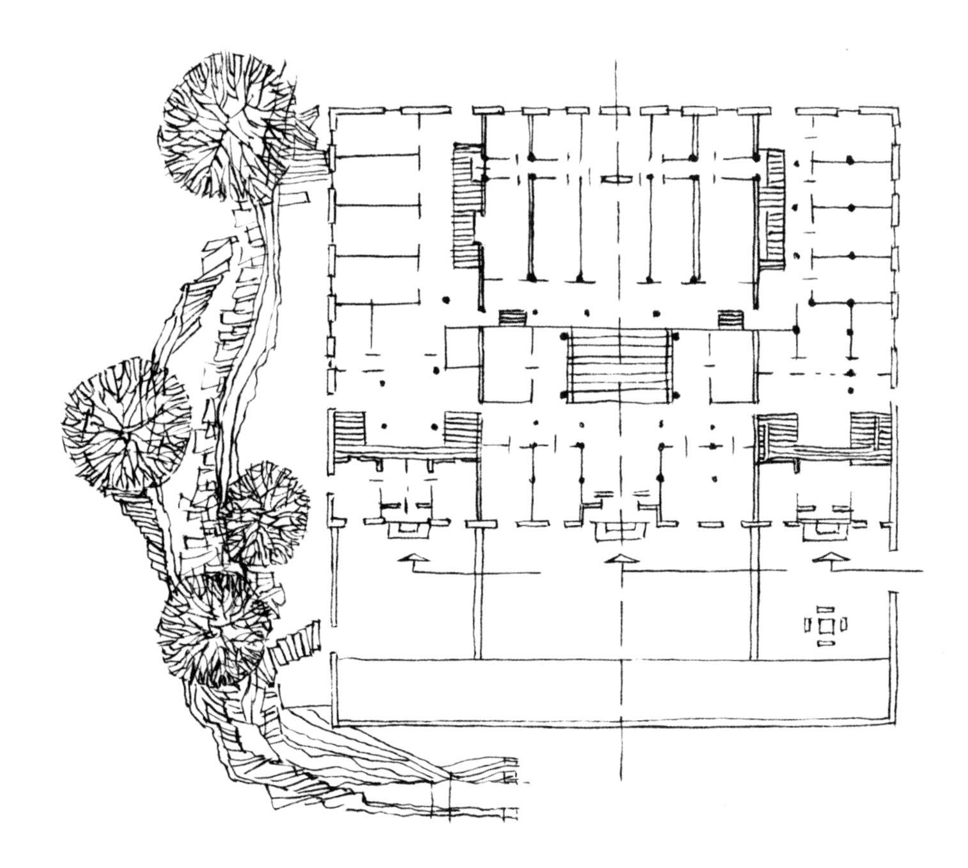

The upper right of the picture above shows the floor plan of the mansion. The yard in front of the main building is the biggest while the rest are considerably smaller. The mansion's yards vary in shape and size, creating a sense of dynamics. Some are square, some longitudinally rectangular, and others horizontally rectangular. The separated yards are interlinked. If conditions allow, the altitudes of the buildings in such a traditional residence gradually rise from front to rear to add a solemn sense. If the buildings of the same architectural complex cannot be constructed along a single central axis due to topography or other reasons, they are divided into several interconnected sections.

13 广东民居
Traditional Residences in Guangdong

沿着山间崎岖的径道继续南下，便到了广东省境内。首先跃入眼帘的是梅县民居。由于这些以泥墙砌筑的民居经不起频繁的强风劲雨的袭击，所以采用联立式，数户一栋，并将房屋高度降低，增加其强度和抵御力量，以防御海风夏雨的侵袭和炎热的气候。在外观上不论楼房或平房，均为硬山。山墙屋檐装饰以及屋脊式样很多，亦很烦琐。建筑用木架结构，一层铺砖，二层则大多用木楼板。村落中，在总体上从前排第二横列房屋起，必须高度递增一尺左右，因此越在后的房屋越高。据说是因为风土关系，但也符合采光通风的要求。这里看到的纵列式楼房是广东客家住宅普遍的形体。广东民居是广东人性格在建筑上的体现，是一扇“南风窗”，和内地建筑风格迥然不同。笔者认为，广东民居更少受“法式”“则例”的限制，是传统民居向现代住宅民居转变的先驱。这种启迪是微妙的，人们会由此感受到广东人开放的性格。

From Fujian, we traveled south over rugged mountain roads to Guangdong Province, where we were greeted by traditional residences featuring clay walls in Meixian County. To withstand frequent winds and rains, several residences are interconnected like townhouses. Moreover, they were built as low as possible to increase their structural stability and enhance their capacity to resist strong winds from the sea and frequent rain and hot weather in summer. Whether being bungalows or multi-story buildings, they all have gable roofs. The gable and eave decorations and roof ridges are varied and intricate. Almost all local residences are made of wood. The first floor is paved with tile, and the second story often has a hardwood floor. In general, the foundations of houses in the village rise by about a foot, row by row from front to rear, making the rear the highest. The practice not only conforms to local customs but also meets the requirements of natural lighting and ventilation. The longitudinal arrangement of houses is a common feature of Hakka residences in Guangdong. The strong Guangdong style of the local traditional residences sharply contrasts homes in inland areas, making them like a window to the architecture of southern China. In my opinion, the rejection of fixed rules and standards in Guangdong's residential architecture—driven by the open-mindedness of Guangdong people—made it a driving force in the wider transition from traditional residences to modern ones.

湘鄂赣民居
Traditional Residences in Hunan, Hubei, and Jiangxi

在湘鄂赣这一有山区、有丘陵，也有冲积平原的地区，春天给万物带来了勃勃生机。天上那些透明的羽毛状云彩与春天融在一起。民居在被隆冬包裹以后，也开始舒展开它的臂膀。这里的民居，大部分为木结构、瓦顶，有的还用重檐，房屋多高大；结构总不脱离横向结构承重、纵向架设檩条的两坡水屋面的基本做法。这里我们看到的湖南湘西土家族苗族自治州民居便是如此。湘鄂赣乡村的民居不论砖墙或木板外墙的房屋常用悬山式。令人向往的是，该地区民居变化极多。总的说来，它的组合是充分结合了该地区天气特点，通过庭院天井间的安排，使住房面对着最佳的朝向与景色，同时使室内外具有密切的联系。房与房之间，采用庭院的檐廊来贯通。

The arrival of spring brings everything back to life in the Hunan-Hubei-Jiangxi region's mountains, hills, and alluvial plains. White clouds float like feathers in the spring sky, and local residences seem to stretch after waking from the lengthy winter. Most traditional residences feature wooden structures and tiled roofs, and some have multiple eaves. Those houses are magnificent and have saddle roofs supported by horizontal girders and longitudinal beams. The residences we saw in Xiangxi Tujia and Miao Autonomous Prefecture, Hunan Province, were fine examples. Whether having brick or wooden exterior walls, rural residences in the Hunan-Hubei-Jiangxi region usually adopt overhanging gable roofs. Local residences are diverse in architectural style, but in general, they are designed to conform to climatic conditions in the region. Skillful arrangement of courtyard buildings often gives bedrooms optimal views. Moreover, indoor and outdoor spaces are closely connected. Different buildings in the compound are linked by roofed corridors.

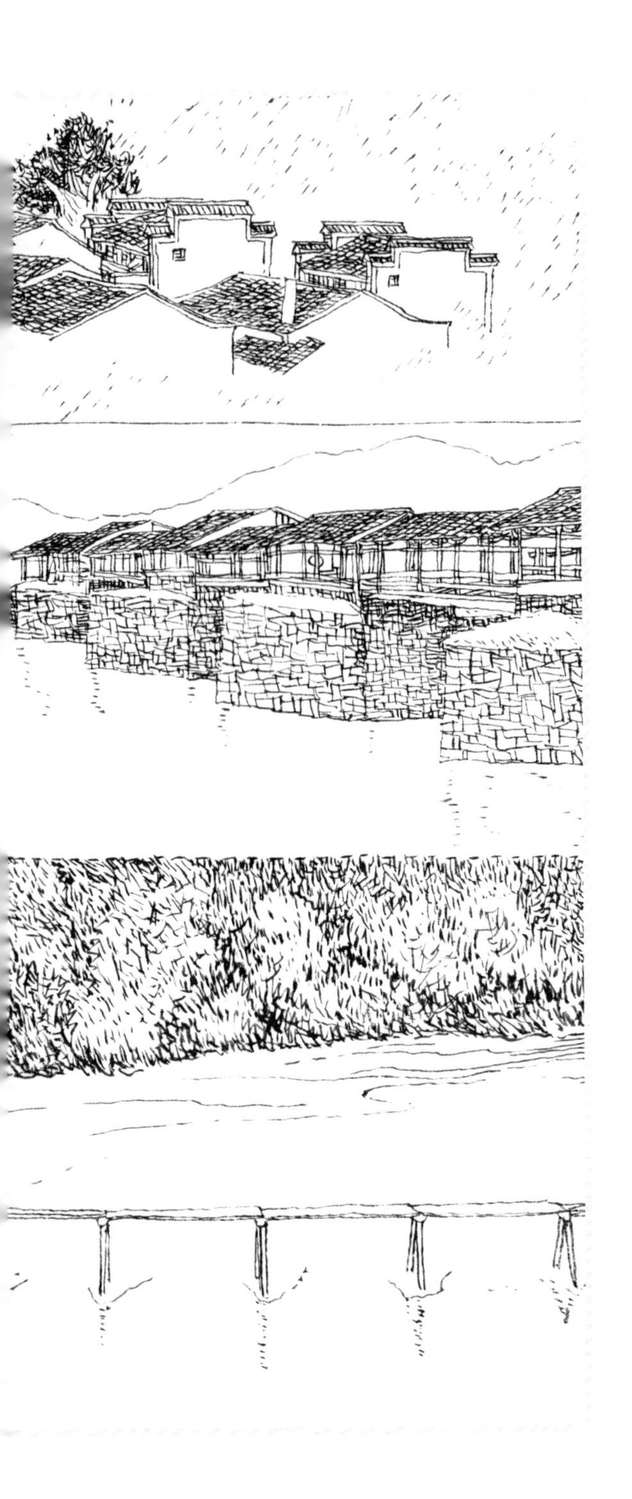

当你欣赏到左侧上方这幅画时，我想，你一定会为那诗一般的绵绵秋雨飘洒在屋顶上的景致而陶醉，也会想象出你走在那木板小桥上时的惬意心情。请到江西婺源去吧，在那里，你将深深体会到“西方建筑长于造型美，中国建筑长于环境美”这句话的含义。

婺源民居具有许多特色。这个县仅延村一个村庄设计独特的古民居就有57栋，布局是“群屋一体”，如遇雨雪天气，可做到穿堂入室，从村头至村尾，衣衫不湿。整个村庄白墙青瓦，古朴而淡泊，其建筑雕刻精巧细致，蔚为可观。

The above picture on the left presents a poetic scene of autumn drizzle dropping on the roofs of old-time houses and the lower one inspires the joy of walking across the wooden bridge. Such experiences are what Jiangxi's Wuyuan County offers. It will make you understand why some have said, "Western architecture is noted for beautiful designs, and Chinese architecture for a beautiful environment."

The residences in Wuyuan are special in many ways. The county's Yancun Village alone has 57 traditional residences with unique flavor. All of them are interconnected. On rainy and snowy days, one can cross the entire village without getting wet. The entire village boasts an impressive expanse of old-style and elegant houses with white walls, black tiled roofs, and exquisite decorative carvings.

上图是湘西民居。这里广泛地使用有顶无墙的出檐。柱廊、过亭等半露天空间，遮蔽雨水与烈日，外观变化多端。木架柱身多外露，与柱间的白粉墙或木板墙配合默契，堪称坦率简朴。

The above picture shows traditional residences in western Hunan. Local buildings, including houses, corridors, and pavilions, widely use overhanging eaves to shelter pedestrians from rain and sun. Their appearances are diverse and dynamic in style. Wooden columns maintain harmony with neighboring white stucco walls or wooden walls, fueling a primitive and straightforward beauty.

左侧上下图均为湖北武汉市某宅。经济条件好的人家为了防雨水、御火灾，常在木构架建筑的外围部分砌筑砖墙，并且超出屋面的高度。这种封火墙或封火山墙成为艺术装饰的重点。而大门上加雨檐或于大门上伸出两个墀形装饰，则是最令人注目的地方，形象丰富。左侧方右上角图为湖南凤凰县街景，这是一个将传统民居保存完好的城镇。现在凤凰县城已成为旅游和拍摄影片的绝好地点。

The two pictures on the left show a residence in Wuhan, Hubei Province. To resist rain and fire, wealthy families often built brick walls outside their wood-structured houses much higher than the roofs. Typically, those brick walls or gables are exquisitely decorated. The gate often has an overhanging eave or two step-shaped decorative components on the top, which are the most eye-catching and dynamic part of the house's facade. The upper right picture shows a street in Fenghuang County, Hunan Province. Fenghuang is an ancient town that has preserved many traditional residences. It is now a popular destination for tourists and photographers.

皖南民居
Traditional Residences in Southern Anhui

皖南建筑以楼房为主，形式多样，但基本上都是木构架瓦房。木构架也基本为穿斗式。右图是某村景。皖南民居的外形变化较大，其木雕和砖雕为全国民居中的佼佼者。尤其值得介绍的是徽州地区至今还有一批保存完好的历史价值和艺术价值均相当高的明代住宅。

Traditional residences in southern Anhui Province are mostly multi-story buildings. Despite their varied architectural styles, most are wood-structured houses with tiled roofs and through-tenon frames. The right picture shows a village in the region. Traditional residences in southern Anhui vary in appearance and are noted throughout China for their outstanding wooden and brick carvings. Huizhou has preserved numerous Ming Dynasty residences with great historical and artistic value.

在艺术处理方面，这类住宅的外部形体简单朴素，可是一入大门，走进院子，却使人印象为之一变。楼上在柱子外侧周围装有华美的木栏杆，栏杆上的花纹有的简洁秀丽，有的复杂细致，在统一中又有若干变化。由于集中使用装饰，使以水平线条为主的雕饰较繁密的栏杆，与上下两层以垂直线条为主、体形比较整齐的木板壁及柳条式的窗棂形成强烈的对比。这也是这类民居最主要且最成功的地方。左图是安徽黟县民居的庭内和室内装饰，属于简洁的实例。

In terms of artistic design, most houses feature simple and unsophisticated facades. But when you enter the compound, you'll be impressed by the gorgeous wooden balustrades outside the pillars on the second floor, of which some are decorated with simple but elegant patterns and others with intricate and exquisite carvings, adding enormous changes to the seemingly unified buildings. The intricate and dense carvings on the horizontal balustrades sharply contrast the neat wooden walls and vertical window lattices on both floors. This is the most prominent feature of local traditional residences and the heart of their strong aesthetics. The left picture shows the simple courtyard and interior decorations of a residence in Yixian County, Anhui Province.

值得提及的是皖南民居华丽的装修雕刻仅限于明代中叶至明末，入清后便逐渐减少，民居艺术中这种作风的兴衰与明代版画的盛衰、人们艺术鉴赏爱好的转移和当地的经济条件状况，以及艺人匠师们的创作水平都有着密切的关系。右图是泾县街景，属于清代民居风格。尽管装饰减少，但韵味犹存。店铺与店铺之间几乎都是层层出挑，加上顶端突出的屋檐，使街道形成一个半封闭的带形空间。

It should be noted that gorgeous decorative carvings were most popular among traditional residences built in southern Anhui during the middle and late Ming Dynasty, and they began to decline during the Qing Dynasty. This shift had close relations to the rise and decline of Ming Dynasty woodeut prints, changes in artistic taste and economic conditions, and the creative capacity of artisans. The right picture shows Qing Dynasty-style residences along a street in Jingxian County, Anhui Province. Although they offer less decorative patterns than their predecessors, the houses remain elegant. Every roadside shop has corridors on each floor. And the overhanging eaves make the street a semi-roofed belt space.

皖南民居的艺术风格是独特的。迷离曲折的空间分隔，规整方正的庭院，雕镂细腻的纹饰，几乎处处都是强烈的对比，但总的效果却又高度和谐。没有离开实用的装饰，没有似是而非的手法，该精则精，该糙则糙，自然直率地表现自己。这里我们看到的是安徽屯溪县街景。这是一幅洋溢着浓厚生活气息的图画。茶馆里，一年四季茶客满座。品茶、听书、闲谈、观赏街景，饶有情趣。尤其是远处的过街廊所形成的立体交叉，构成了一个灰空间，使行人感到步移景异。

Traditional residences in southern Anhui boast a unique aesthetic style. Labyrinthine spatial partitions, square courtyards, and exquisite decorative carvings abound. Despite many strongly contrasting elements, the entire residences still radiate harmony. They are immune to both impractical decoration and pompous technique. Exquisiteness and roughness coexist perfectly, and all parts of the residence are displayed straightforwardly. This picture shows a lively street in Tunxi County, Anhui Province. The teahouse is packed with customers all year round. They sip tea, read, and chat while enjoying the streetscapes. The covered corridor in the distance forms a gray space, allowing pedestrians to enjoy different views as they walk down the street.

晋陕民居 Traditional Residences in Shanxi and Shaanxi

从明代起，许多山西人便外出经商，致富返乡后，在自己故里纷纷大兴土木。他们不但要舒适，而且要华丽、坚固（便于防卫）。大户人家几乎全是灰砖高砌、居室密集的深宅大院。这类住宅的临街大门甚是堂皇，有漂亮的门楼。大门内为雕砌的砖影壁，进入后忽然现出垂花门（即内宅门），庭院布局为二进或三进的四合院。右图是山西省太谷县某宅。

晋陕民居房屋密度大，为防风沙与日晒多采用窄天井，且庭院内正房和厢房多有廊。乡间大宅常有一座小方形砖楼供瞭望用，即所谓“看家楼”。平时登楼遥望田野，的确是很惬意的乐事。房顶的形式有“人”字顶，在前廊的额枋等处，常有彩画雕刻，富丽可观。

Since the Ming Dynasty, many Shanxi people have ventured outside to do business. After amassing wealth, they returned home and built grand residences. Their houses were not only cozy but also imposing and fortified (for defensive purposes). The residences of wealthy families are mostly mansions featuring tall gray brick walls and densely distributed buildings. They usually have resplendent and gorgeous gate towers facing the street, inside which is a screen wall with exquisite brick carvings. Such a mansion usually consists of two or three courtyards. A festoon gate separates the outer yard and the inner yard. The right picture shows a mansion in Taigu County, Shanxi Province.

Traditional residences in Shanxi and Shaanxi feature a high density of buildings. To resist sandstorms and the strong sun, they mostly adopt small courtyards, and the main halls and side halls have roofed corridors. The mansions in the countryside often have a square brick watchtower, which is called "Home-watching Tower." It is pleasant to overlook the nearby fields from atop the watchtower. Most houses feature saddle roofs, and the architraves of the front corridor are often decorated with gorgeously painted carvings.

晋陕一带普通人家常有一面坡屋顶，也有一面坡屋顶组成的三合或四合院。房顶向院内倾斜，包括大门与倒座也用一面坡屋顶向院内排水。这样的建筑群外墙很高，有安全感。左侧左上图为陕西临潼某宅，左图右下部分为陕西西安某宅。

由于夯土做成的外墙比较高耸，有时在墙身上都装薄砖两排，增加墙身的强度以防止雨雪侵蚀墙面。同时在外观上可配合上部出檐，增加水平印象，使平坦的墙面发生变化。这一地区院庭的形状均南北狭长，东西屋的山墙将堂屋东西次间遮住。这种形式在华北、西北地区较普遍。有趣的是与南方东西长、南北狭的院子恰恰相反。

The residences of ordinary households in Shanxi and Shaanxi provinces often feature single sloped roofs. Some have three or four courtyards consisting of houses with single sloped roofs. The roofs, including those of the gate tower and north-facing houses, are sloped toward the courtyards so rainwater flows inside. The tall exterior walls foster a sense of security for dwellers. The upper left of the left picture is a residence in Lintong, Shaanxi, and the lower right is a residence in Xi'an, Shaanxi.

The tall exterior walls are made of rammed earth. Some are inlaid with two lines of thin bricks to enhance their solidity and prevent erosion from rain and snow. Moreover, the overhanging eaves on the top intensify the horizontal visual effect and add dynamics to the flat wall. The courtyards of residences in the region are longer north to south, and the gables of the houses in the east and west wings shelter the side rooms of the main hall. This architectural style is popular in northern and northwestern China. Interestingly, courtyards in southern China are just the opposite: longer east to west.

17 干栏式民居 Stilt Houses

位于我国西南的云南省、贵州省以其风景奇丽而吸引四方来客，这里的民居也风格各异。境内的少数民族包括傣族、景颇族、佤族、哈尼族和水族等都有使用“干栏式”住宅的。这种建筑形式历史悠久，《魏书·僚传》说“依树积木，以居其上，名曰干栏”。干栏式住宅的特点是用竹或木为柱梁搭成小楼，上层住人，下层作牲畜圈或储存杂物之用。《旧唐书》解释：“人并楼居，登梯而上，号为干栏。”干栏式民居又分高楼式和低楼式，即是对下层透空柱梁空间的高度而言。右图是贵州水族民居，为高楼式干栏，上层为卧室，中层为起居室，下层为仓库和牲畜圈。

Yunnan and Guizhou provinces in southwestern China attract tourists from around the world with picturesque landscapes. The traditional residences in the region vary in style. Local people from ethnic minorities such as the Dai, Jingpo, Va, Hani and Shui usually dwell in stilt houses, a type of residence that originated a long time ago. Stilt houses were defined as "dwellings built with wood on trees" in *Biography of Liao in the Book of Wei*. Today, stilt houses refer to multi-story residences supported by bamboo or wooden stilts beneath. The upper floor is for people to live, and the lower floor is used as the stable and storage room. The *Old Book of Tang* explained that "people live on the upper floor which is accessible by a stairway, and this kind of residence is called stilt house." Based on the height of stilt pillars beneath, stilt houses are categorized into tall houses and low ones. The right picture shows a tall stilt house built by Shui people in Guizhou Province. The uppermost floor is the bedroom, the middle floor is the living room, and the lowest floor serves as a storage room and stable.

干栏式住宅尽管室内较暗，但出檐深远，遮住阳光的辐射，外廊也对此做了补救。对于多雨潮湿的地面有隔离作用，通风较好，适用于当地气候；另外，在人烟稀少的地区，还可以防范野兽伤害。右图是贵州镇宁布依苗族自治县某宅。笔者曾在这个房子里住过。由于是老屋，从地板的缝隙可以清楚地看到楼下的牛圈。虽然有蚊虫轮番进行“空袭”，但旅途的疲劳使我在牛粪的气味中鼾然入睡。黎明时分，一个沉闷的响声使我从梦中惊醒，睡眼惺忪中这声音又使地板轻颤一次。呵！原来楼下的牛已经在吃草了。

Because their outstretched eaves block the sunlight, stilt houses are relatively dark indoors even in the daytime. However, the spacious outer corridors make up for this deficiency. The region features frequent rainfall. The stilt pillars enable the floor to be isolated from wet ground and facilitate ventilation. Such a design conforms to local climatic conditions. It also protects people from intrusion of wild animals in relatively unpopulated areas. The right picture shows an old stilt house in Zhenning Buyi and Miao Autonomous County, Guizhou Province. I was able to stay in the house for several days. Through cracks in the floor, I could clearly see the stable beneath. Despite insect bites and the odor of cow manure, I quickly fell deeply asleep because I was so exhausted from the trip. At dawn, I was awoken by a loud sound that even shook the floor slightly. Dazed, I turned my drowsy eyes downward to see the cattle downstairs had already begun chewing their straw breakfast.

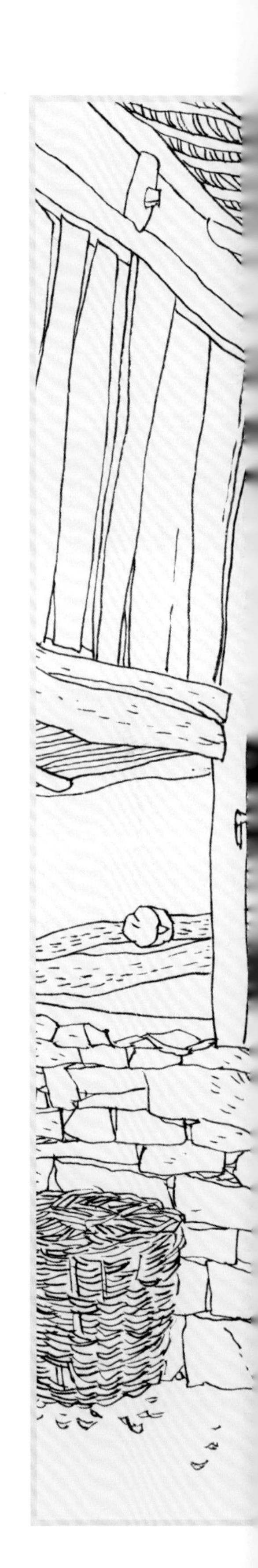

这里我们看到的是云南景颇族外廊式民居，这是一种低楼式干栏。据云南祥云大波那村木铜棺墓出土的随葬青铜器证明，大约在公元前 400 年的战国时期，就已有这种长脊短檐倒梯形屋面的干栏建筑。这是因为古人常在山尖屋脊下悬挂牛头等猎物，以表示房主人的勇敢与勤劳。图中的民居，楼下已不关牲畜，而作仓库使用。一般都设单独厨房，这样也改善了居室内的卫生条件。海南黎族船形民居也是一种矮干栏建筑，有的草顶垂下紧贴两侧，竹笆形成筒状，有的在屋山一头用草顶做成半球形以作谷仓。因外形很像一艘船，故名船形民居。

This is a residence of the Jingbo people in Yunnan Province, viewed from different angles. It is a low stilt house with exterior corridors. Bronze ware unearthed from a tomb with wooden and bronze coffins in Dabona Village in Yunnan's Xiangyun County showed that this kind of stilt house with inverted trapezoidal roofs, long ridges and short eaves dates back to at least the Warring States Period (475-221 B.C.). In ancient times, house owners often hung the heads of the animals as trophies under the eaves to manifest their bravery and diligence. The lower floor of the residence pictured no longer serves as a stable, but a warehouse. Such houses usually are equipped with an individual kitchen to improve the sanitary conditions indoors. The boat-shaped residences of the Li people in Hainan Province are also a kind of low stilt houses. Some have drooping thatched roofs, some have tube-shaped bamboo fences, and some have a hemispheric barn covered with thatched roof connecting to one side of the gables. Such residences resemble boats, hence the name "boat-shaped houses."

一提到傣族，人们很自然地就会联想到西双版纳的亚热带原始森林，傣家姑娘的孔雀舞，傣乡泼水节，还有那凤尾竹下的小竹楼。竹楼也是一种干栏式建筑，在德宏和西双版纳分布最广。现在的“竹楼”所用材料主要是木料。楼上有前廊、晒架、内室三部分。前廊是民居活动之处。内室一般为一通间，无床，辈分晚者居外，辈分长者居内，房屋多数为歇山式屋顶。这里最美的景是晚上，皎洁的月光下，凤尾竹轻柔得像绿色的雾一般，一切都那么温婉可人，令人意念回旋。

When the Dai ethnic group is mentioned, people naturally think of subtropical primitive forests in Xishuangbanna, the peacock dance performed by Dai girls, the Water-splashing Festival, and bamboo houses hidden in fernleaf hedge bamboo groves. Dai bamboo houses are also a kind of stilt residences popular in Dehong and Xishuangbanna, Yunnan Province. However, today the Dai people mainly use timber instead of bamboo. The upper floor of such a residence consists of a front corridor, a shining platform, and a bedroom. The spacious front corridor functions like a living room. The bedroom is spacious but actually lacks beds. The younger members of the family sleep on the floor close to the door, while the seniors sleep inside. Most of the local residences feature hip and gable roofs. The houses look beautiful, especially at night. Bathed in bright moonlight, fernleaf hedge bamboo sways in a gentle breeze, creating an illusion of green mist. Everything is adorable, elegant, and inspiring.

维吾尔族民居
Uygur Residences

从乌鲁木齐东进穿过一望无际的戈壁沙漠，方能到达吐鲁番——充满活力的绿洲。吐鲁番盆地全年基本无雨，但每年的春季则要“下土”几天。那时，到处都是漫漫黄土，遮天敝日，维吾尔族的土坯平顶住宅也淹没在这黄色苍茫之中。这里除生活富裕的人家用砖修建房屋之外，一般人家都就地取材，用土坯修造，依地形组合为院落式住宅。在布局上，院子周围以平房和楼房相互穿插。维吾尔族建筑空间开敞，形体错落，灵活多变，用土坯花墙、拱门等划分空间。上方左图是吐鲁番葡萄沟某宅，上方右图是土峪沟某伊斯兰风格的建筑。

Setting out from Urumqi, capital of northwestern China's Xinjiang Uygur Autonomous Region, the trip to the oasis of Turpan requires crossing a boundless desert. The Turpan Basin sees little rainfall all year round and suffers frequent sandstorms in spring. When a sandstorm strikes, local Uygur earth houses with flat roofs become submerged in yellowish sand and dirt. Except for a few wealthy families who live in brick houses, most homes are built of earth and arranged according to the terrain to form compounds. In terms of layout, bungalows and multi-story buildings are randomly distributed. Uygur residences feature spacious and dynamic layouts and use decorative earth walls and arched doors to partition different spaces. The left of the picture above shows a residence in Grape Valley, Turpan, and the right shows an Islamic building in Tuyugou.

吐鲁番盛产葡萄，连街道上也有葡萄长廊。这里夏季白天气温高达 47℃，而夜间只有 20℃；白天穿背心裤衩，夜里却得盖棉被。因此用生土建筑的墙体特别厚。到小巷中去观光，步行在民居之中，会感到意味无穷，忽而一个跨度很大的土拱，上面是院落，下面是道路，仿佛是现代的立交桥；忽而街道变窄，人从隧道般带有土拱的长长的小巷中走过。维吾尔族民居以前室和后室相结合，附以厨房、马厩等。由于大陆性气候非常明显，气温变化大，一般不开侧窗，只开前窗，或自天窗采光。吐鲁番盆地每年四、五月都要刮二十多天十级以上的大风，所以民居保持一定的密度，庭院也留得不是很大。

Turpan abounds in grapes, and corridors covered in grapevines can be found in almost every street and lane. The daytime temperature can reach as high as 47 degrees Celsius in summer and drop to 20 degrees Celsius at night. Locals wear vests and pants in the daytime and cover themselves with quilts at night. For this reason, local earth houses adopt thick walls to resist heat in the daytime and keep warmth at night. Many lanes have impressive residences on both sides, and huge earth arches can catch you by surprise. Above one arch was a residential compound and beneath it a path, so it functions like a modern overpass. Then, the street suddenly becomes narrow and pedestrians navigate a long tunnel-like lane with earth arches. A Uygur residence consists of a front house and a rear house as well as other facilities like kitchen and stable. Due to the typical continental climate and large temperature difference in the region, local residences usually have no side windows, but use front windows and skylights to allow sunlight to penetrate. In April and May, Turpan often sees more than 20 windy days. To resist winds, local residences are densely distributed and lack large courtyards.

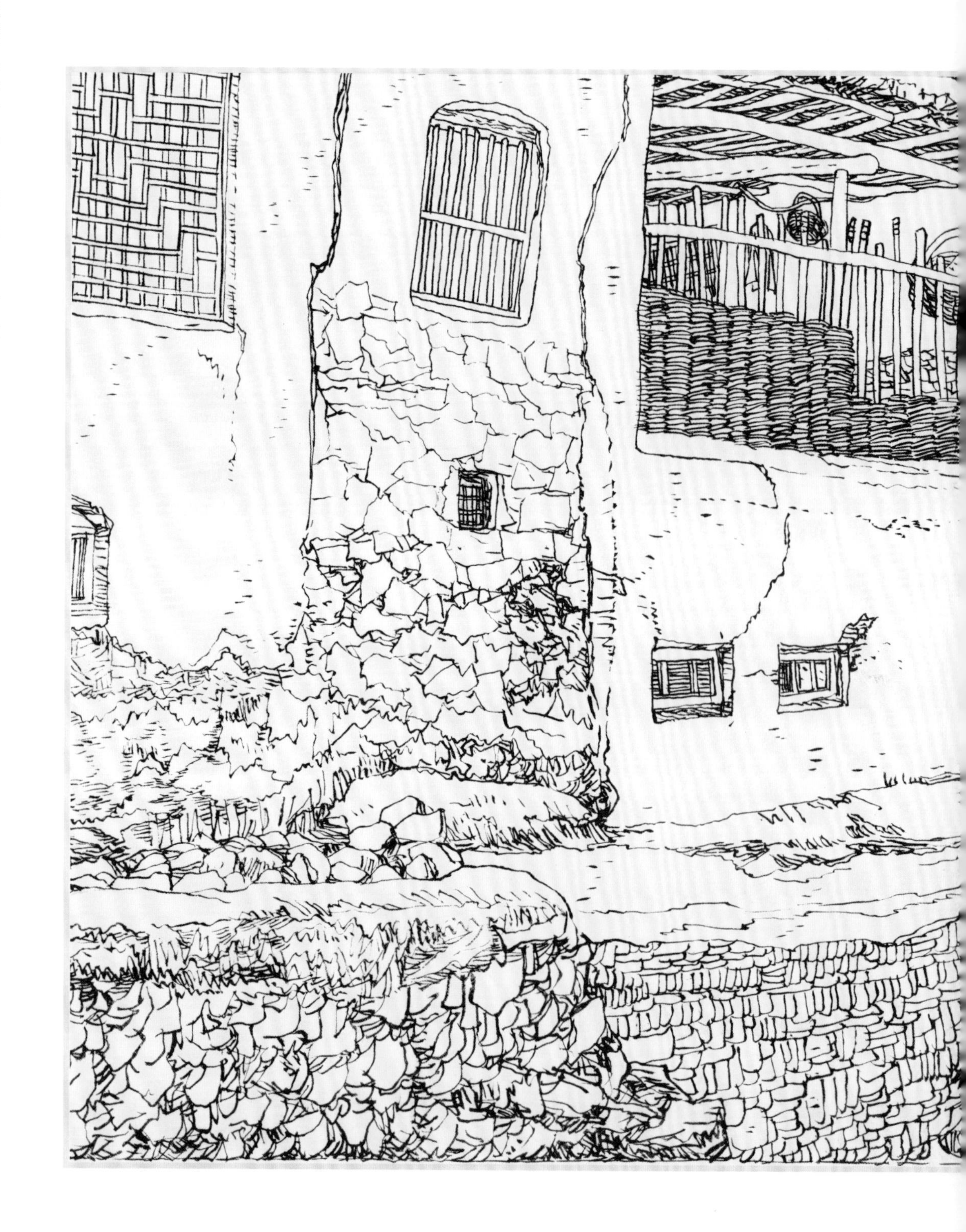

上图是库车某宅。我们可以看出维吾尔族民居的灵活风格，开窗不拘成法，大小随需要而定，窗棂花格多样，房屋外形变化丰富，很有特点。

The picture above shows a residence in Kucha that demonstrates the flexibility of Uygur houses. There are no fixed rules for opening direction or size of windows, so all are designed to meet actual needs. Local houses vary in window lattice patterns and architectural shapes.

从喀什沿着塔克拉玛干大沙漠边缘南下，经莎车到和田，这一路给人印象最深的是土多。只要出趟门就变成了“土人”。当地的朋友开玩笑说，假如你骑驴摔下来就没了——被土埋起来了。这里四季气候干燥，所以很适合生土建筑。由于生土房子经不起较大降雨，所以富有人家仍用砖砌住宅。拱廊、墙面、壁龛、火炉与密肋、天花等处，雕饰精致；又因信奉伊斯兰教，故多用绿色来装点，但普通人家室内装饰还是比较简单的。维吾尔族人习惯在墙上挂着美丽的壁毯以作装饰。右图是喀什某宅，因坍塌了一部分，我们可以看到剖面。

I was impressed by the volume of dirt and sand accompanying the journey from Kashgar to Yarkand and then Hotan along the edge of the Taklimakan Desert. One can become a "dirt person" after staying outdoors for a while. A local joked that if you fall from your donkey, you'll get buried in dirt. It is dry all year round in the region, so earthen residences are suitable. However, earth houses are susceptible to heavy rains, so wealthy families still tend to use brick. Local houses have exquisite carvings on arched corridors, interior walls, niches, stoves, and ceilings. Due to their Islamic belief, locals use green as the main tone to decorate their houses. However, interior decorations in ordinary households are relatively simple. Uygur people habitually decorate their interior walls with beautiful tapestries. The right picture shows a residence in Kashgar with a collapsed part enabling a cross-sectional view.

贵州石板房
Stone Houses in Guizhou

贵州位于云南高原东部，山丘隆起，覆土较少，遍地岩石。当地人就地取材，采石建房，有的甚至利用平整的山岩作为民居墙体，房屋构架为木材制作的穿斗式。屋脊坡面用薄层石灰岩做瓦，巧妙地解决了屋脊漏水的问题。铺地用的是石板，楼板也是石片，水缸则用大块石板拼成四方体，牲口槽也用石块凿成。石头的色彩在灰调子中呈现出白、蓝灰、浅土红等色相，既简单又复杂，既细致又宏大。贵阳花溪石板哨，镇宁、安顺一带以及陕西安康等地都有石板房。有机会去黄果树瀑布时，只需步行半个多小时就能游览镇宁石头寨。

Located on the eastern Yunnan-Guizhou Plateau, the province of Guizhou features rolling mountains and a very limited amount of arable land, with most territory covered with rocks. Thus, locals use stones to build their houses. Some residences use natural cliffs as walls. Most traditional houses adopt a through-tenon wooden framework. Their sloped roofs are covered with thin limestone planks to skillfully prevent leakage on rainy days. The floors are also paved with stone planks, and even water vats and cattle troughs are made of stone. With gray as the main tone, the stone houses also feature colors like white, blue, and red, which makes them look simple but dynamic, exquisite but magnificent. Such stone houses are found in Shibanshao of Huaxi District in Guiyang City and Zhenning County in Anshun City as well as Ankang City, Shaanxi Province. It takes only a half hour to walk from famous Huangguoshu Waterfall to Zhenning Stone Village.

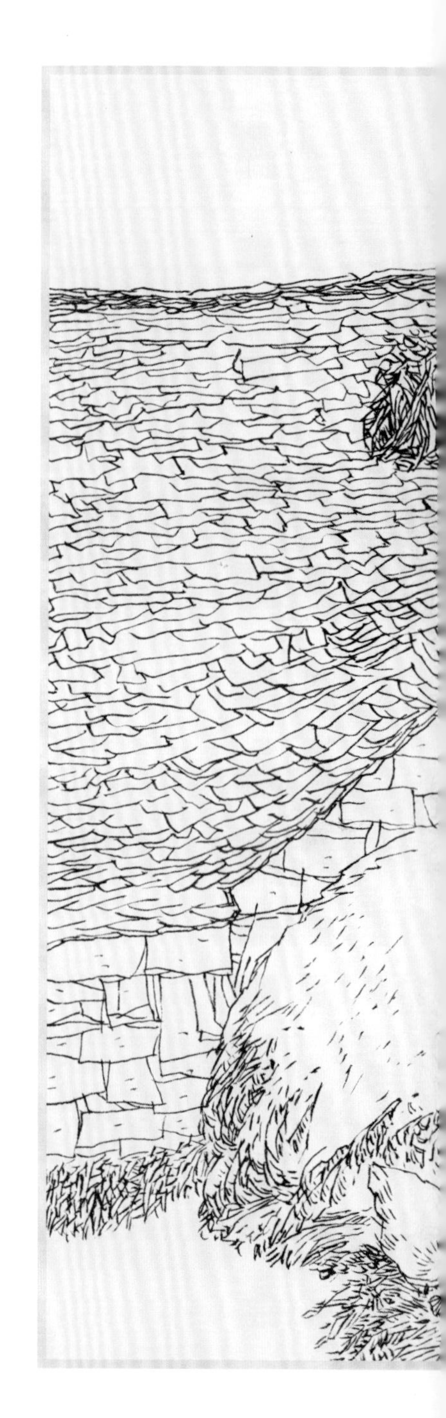

20 井干式民居 Log Houses

井干式和干栏式一样，也是我国古老的建筑形式之一。从云南石寨山出土的贮贝器、铜器的纹样上，可以看到井干式房屋的样式，证明在汉代已经有了“井干”这种构造方法。汉武帝时造的井干楼是很高的，被描绘为“襻井干而未半，目旋转而意迷”。这里我们且不去讨论文学描写的夸张性，仅从建筑的角度来分析。井干式房屋的外墙和内墙都是用去皮圆木或方木层层垛起，木楞接触面做成深槽，利于叠紧稳固并防水。墙角处交叉相接，中间隔墙的木楞也交叉外露，叠积的圆木粗率地暴露在外，不饰油漆，因形如井口，故称为“井干式”，也有叫作“木楞房”或“垛木房”的。张衡《西京赋》里“井干叠而百层”，形象地描述了“井干式”建筑。目前我国主要在东北、新疆和云南等林区有这种建筑。房顶基本为悬山式，有的在缝处抹泥以防风寒，房顶有草顶、树皮顶，比较有代表性的是木片顶。井干式民居以大分散小集中的形式组成村落，目的是为了防火。我们这里看到的是云南永宁纳西族一处村寨。

Like stilt houses, log houses are also one of the oldest genres of Chinese residences. The shell containers and bronze ware unearthed from Shizhai Mountain in Yunnan Province showed that log houses already emerged as early as the Han Dynasty. Emperor Wu of Han once built a tall log tower, and according to historical records, “one can feel dazed and exhausted after climbing only half of the tower.” The description may have used the literary technique of exaggeration, but we can look closer at such houses from the perspective of architecture. Both exterior and interior walls of log houses are laid with stripped logs or rectangular timber, which are closely interconnected by mortise and tenon joints to prevent water leakage. At the corners, logs are intersected as are those of

the partitioning wall in the middle. All logs of the houses are naked and unpainted. Such log houses look like water wells, hence the alternate name "Jingan" (meaning "well-shaped houses") or "timber-piled houses." In *Ode to the West Capital*, Zhang Heng, a Chinese scholar and scientist in the Han Dynasty, vividly described log houses as "buildings piled with 100 layers of logs." Today, log houses are mainly found in forest areas in northeastern China as well as Xinjiang Uygur Autonomous Region in northwestern China and Yunnan Province in southwestern China. They usually feature overhanging gable roofs, and some seal the cracks with clay to resist wind and stay warm. Some are roofed with thatch or bark, but the most representative is a wooden plank roof. Log houses are sparsely distributed in a village so that if one catches fire, it will not spread to others. The picture shows a Naxi village in Yongning, Yunnan Province.

21 白族民居
Bai-style Houses

白族民居以绚丽精致、绰约多姿而著称，具有浓郁的民族特色。为适应风大、地震多的环境特点，平面布局上的典型形式是“三坊一照壁”及“四合五天井”。“三坊一照壁”是由三合院与一个美丽的照壁组成，这种形式数量较多，是主要形式。右图所示为“四合五天井”，就是四合院共有大小五个天井，坊即一栋三开间二层的房子。白族人常说：“正房要有靠山，才坐得起人家。”故建筑主轴线的后端正对附近一个吉利的山峦，而最忌背对着山沟或空旷之处。

The traditional houses of the Bai people are noted for gorgeous, exquisite and elegant designs and strong ethnic flavor. Because the region suffers strong wind and frequent earthquakes, local residences are typically in the form of "a compound with houses in three directions and an exquisite screen wall in the other" (most popular) or "a compound comprising five smaller courtyards." (pictured). The main structure is a two-story house with three rooms on each floor. Bai people believe that "a blessed residence should recline against a hill." Therefore, the rear end of a residence's central axis is often against a hill and should avoid reclining against a valley or an empty space.

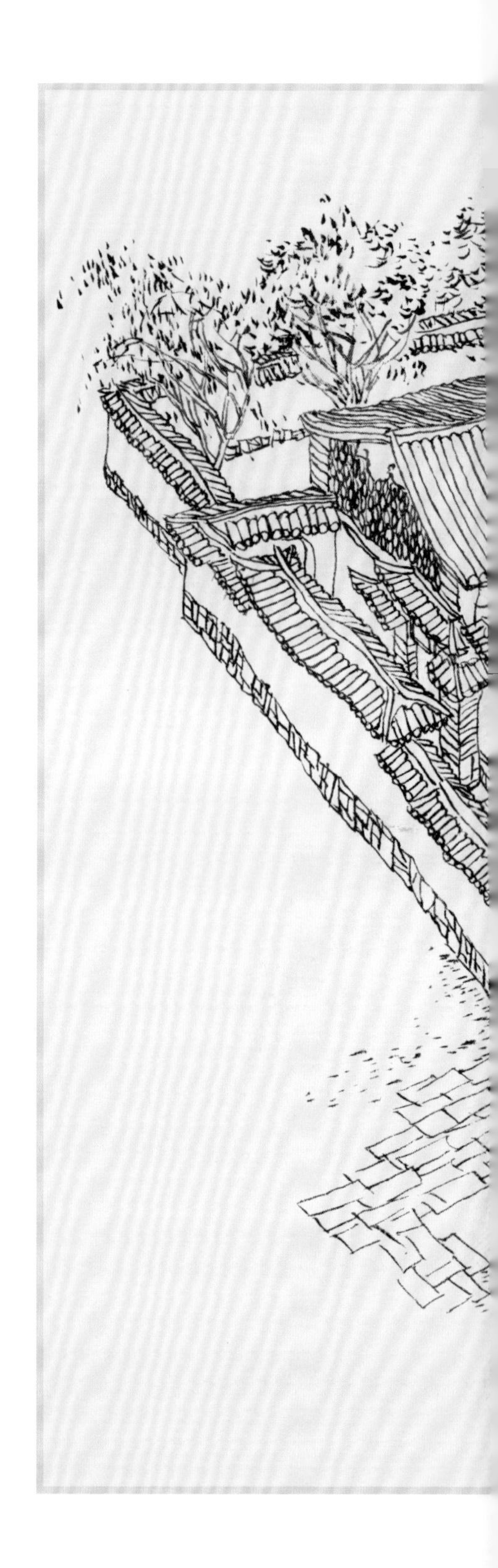

22 西北民居
Traditional Residences in Northwestern China

民居的构造千差万别，各具胜境。广袤的大西北，不乏动人的美景。在六盘山脉东部是苍莽雄浑的陇东高原，海拔1000米以上，黄土层厚达150米。右图左上部分是甘肃定西民居，厚重朴实，与陇东高原的地貌十分和谐。右图右下部分是青海东部民居“庄巢”。庄巢民居四面有高厚的土筑外墙，里面用木构架承重，黄土作屋面，内有廊檐。就像当地人那样厚道纯真，倾注着对黄土地的热爱。

Despite varying architectural structures and styles, each kind of residence has its own unique beauty. The vast land of northwestern China boasts beautiful scenery. East of the Liupan Mountains is the magnificent Longdong Plateau at an altitude above 1,000 meters and loess soil as thick as 150 meters. The *upper left* of the right picture shows an imposing and unsophisticated residence in Dingxi City, Gansu Province, which is in harmony with the landforms on the Longdong Plateau. *The lower right* of the picture shows Zhuangchao (literally "Village Nest"), a kind of traditional residence in eastern Qinghai Province. This kind of residence features tall and thick earth exterior walls supported by wooden framework as well as corridors with eaves. The simple and straightforward architectural style mirrors the character of local people and their affection for the loess land.

下篇
Chapter 2

村镇面貌

Villages and Towns

村镇选址
Site Selection for Villages and Towns

选址定位是件大事，被称作相地。相地就是选择地段，进行环境设计，使房屋在优美的自然环境中落位。相地理论的许多内容都与“堪舆学”有关。老子曰：“人法地，地法天，天法道，道法自然。”堪舆学主要利用阴阳五行八卦的道理解释自然。正如一部电视机放在室内的不同位置，收看效果不同一样。同理，人在不同的自然环境中居住，心情也会产生差异。所以堪舆学在选择有利的地理、气候、水文、风土条件方面是有一定道理的。相地之美在于巧，是相地灵活性与环境适应性的结合，“园基不拘方向，地势自有高低”（《园冶·相地》）。

Site selection, also called "land physiognomy," is crucial for construction projects. It refers to the process of selecting a location to build houses and often involves seeking a beautiful natural environment. Many theories related to site selection came from geomancy. Lao-Tzu, the founder of Taoism, once said, "Humans are abstracted from the world, the world from nature, nature from the Way, and the Way from what is beneath abstraction." Geomancy seeks to interpret nature with Taoist theories such as *yin* and *yang*, the Five Elements, and the Eight Diagrams. A TV set, if placed in different parts of the room, can create different visual effects. Similarly, a person's mood can change according to the natural environment where he or she dwells. Thus, it makes sense that people use geomantic theories to choose locations with preferential geographical, climatic, and hydrological conditions. Site selection requires special skills and focuses on both flexibility and adaptability of the environment. "The site for a garden doesn't need to havea fixed direction, but the terrain should have ups and downs" (*Geomantic Theory for Gardening*).

相地之后，须精心地立基，使建筑的布局合乎体理，“宜亭斯亭，宜榭斯榭”（《园冶·兴造论》）。立基合宜得体，便构成了有机的序列，一般村镇多选址在大路河道的附近或傍山近水，朝向良好，避风防洪，地势干爽的平面，或山岗缓坡、台地等处。

沿河村镇基本形式是一面临水或背山面水，建筑沿河道伸展。临水设码头联系水陆交通。建筑多在河流北岸，以取得良好的通风日照。而背山面水，村镇既得近水之便，又地势高爽，可避免河道涨水时被淹，一般在山南，夏日接纳南风，冬日接纳日照，并利用山作为屏障遮挡北来的寒风。上图是苏州光福镇。

After the site is selected, the foundation is carefully laid and architecture designed. "Build pavilions and waterside houses in suitable locations," exhorts *Construction Theory for Gardening.* A fitting foundation leads to an organized layout of buildings. Typically, planners chose to build a village or town near a road or watercourse or close to mountains and rivers, in a location with an unblocked view yet resistant to winds and floods such as warm and dry flatlands, gentle hillsides, and plateaus.

A riverside village or town usually faces the river on one side and reclines against a mountain on the other side, with its buildings lined up along the watercourse. Often a ferry or wharf connects river and land transport. Usually, residences are located on the north bank of the river to ensure ideal ventilation and sunshine. Facing the river and leaning against the mountain, such villages and towns enjoy the convenience of the river while staying free from floods due to relatively high altitude on the hillsides. They typically sit on the south slopes of mountains, which enables exposure to cool southern winds in summer and warm sunshine in winter. Moreover, the mountains shelter them from cold northern winds during winter. The picture above shows Guangfu Town, Suzhou City.

王勃的《滕王阁序》中说："鹤汀凫渚，穷岛屿之萦回。桂殿兰宫，即冈峦之体势。"山区民居的布局，最准确地表现出即冈峦之体势的风格。东南、西南山区，村镇往往是竖向布置，形成层层交叠的构图风格，秀丽婉约而灵活；北方村镇常选平整的地形，街道宽敞，风格严整雄壮而开阔。有些自然村，在距离河道不十分远时，往往会人工挖掘河槽，将河水引入适宜地点，村落就围绕河槽尽头布置开来。尽管河槽末端水流不畅，水质较差，但村民可以取得生活用水或农业灌溉用水。在南方，水土流失较少，故河道的水上交通直达村落，方便了村里的居民，也不必经常清挖河道。

"Circling are the wild ducks on the sand-bars," wrote Tang Dynasty poet Wang Bo in *Preface to Tengwang* Pavilion. "Cassia-wood courts and magnolia-wood halls rise and fall like mountain ranges." The layout of residences in a mountain village can accurately demonstrate the terrain of hills. Villages and towns in mountainous areas of southeastern and southwestern China usually stretch up the hillsides, with residences distributed layer by layer, creating an elegant and flexible sense. Those in northern China are often located on a vast, unblocked flatland with wide streets and magnificent buildings. For villages close to watercourses, locals dug canals to draw water to a suitable location, with their residences built around a pond at the end of such a canal. Though the water from the pond is usually unfit to drink, it is convenient for washing and irrigation. In southern China, where water and soil erosion is negligible, natural rivers flow through some villages and towns, so local residents not only enjoy convenient water transport, but also hardly need to dredge watercourses.

02 街道尺度 Streets and Lanes

民居在传统场镇中以街的形式所构筑的生活环境，其尺度感反映了人们特定的生活方式。这种曲折多变的街道，客观上为人们提供了生活交流的便利条件。合理布局的建筑都能给人带来乐趣，维持着良好的生活环境。北方的老城镇中，许多小巷很窄，不足一米宽，如山西平遥和江苏徐州等地都有被称为“一人巷”或“一线天”的小街道。江南水系发达，常以船作为交通工具，河道形成水街。街道上车辆不多，街面较窄。居民在房上伸出竹竿到对面的房子，即可晾晒衣物。中国传统民居的街道在处理上，突出了“人”这个主题。窄窄的街道、错落有致的建筑立面，步行其中让人备感亲切。

The layout of the streets and lanes as well as residences in a traditional village or town testifies to the features of the local lifestyle. Essentially, zigzagging streets and lanes facilitate communication in daily life. Rationally distributed buildings not only please the eye, but also facilitate a sound living environment. In old towns of northern China, many lanes and alleys are narrow, less than a meter in width. For example, some narrow lanes are figuratively called “one-man alley” or “a crack of sky” in places like Pingyao in Shanxi Province and Xuzhou in Jiangsu Province. Towns in southern China boast abundant river courses, so locals often use boats for transport through canal traffic. For this reason, their streets are relatively narrow, with fewer vehicles. Residents on the second floor often stretch a bamboo stick to the house on the opposite side of the street to dry washed clothes. The design of streets and lanes in traditional Chinese villages and towns is people-centered. It is a great pleasure to roam down such a narrow lane while enjoying beautiful and elegant buildings on both sides.

03 死胡同 Dead Ends

北方的街道纵横交错，一览而尽。而江南地促，故巷道迂回弯曲，若不可测。这种城镇面貌，与江南水乡的自然环境取得审美上的协调。在村庄或城镇街坊内部，巷道或弄道多呈不太规则的形式。自然形成的曲曲折折的巷道虽然比较狭窄，却能避免过境交通，保持宅区内部的安宁静谧，同时使空间变化丰富，使人一进弄堂，犹如进入家园。现代居住区内部道路常采用尽端式的死胡同方法，其理相同。

Crisscrossing streets in northern China are often so straight that you can see all the way down. However, the more bumpy terrain in southern China makes village streets frequently form a labyrinth. This design aesthetically coordinates with the idyllic natural environment of water towns in southern China. Inside such villages and towns, lanes and alleys are often irregular. The naturally formed zigzagging lanes are narrow, but this helps limit the traffic and maintain the tranquility of residences on both sides. Moreover, it adds dynamics to the environment, making it feel like home when walking inside such an alley. For the same reason, internal roads in modern residential areas are often designed with dead ends.

04 广场
Public Squares

广场分为交通广场、集市广场、入口广场、水上广场、水源广场等。陆上交通的五叉、三叉、十字路口及巷子的转折点，常有一个小广场作为交通缓冲和人群流动停留之处，就是交通广场。广场在旧城镇中很少合理规划，大多是从某些功能出发，在城镇发展过程中自然形成。一般都是不规则平面，面积也不是很大。右图是江苏吴江同里镇的一个交通广场。

In terms of function and location, public squares can be categorized into traffic squares, market squares, entrance squares, water squares, and water pond squares. There is usually a small square at crossroads of three to five-forked roads or at turning points of a lane to buffer the traffic and facilitate flow of pedestrians. Such squares are called traffic squares. In old towns, few squares were pre-planned, but formed naturally with a certain function alongside the evolution of the towns. The squares are usually irregular in shape and small in size. The right picture shows a traffic square in Tongli Town of Wujiang, Jiangsu Province.

在村镇入口或热闹集市的附近，往往开扩一部分水面，形成水上广场，作为流动船只或停泊船只之用。较大的村镇往往还建有水上戏台。水上戏台使这种水上广场变为文化娱乐性场所。左图是苏州地区的一个水上广场，小船从水上划过去时，便划出轻轻的影和曲曲的波。有月的夜晚，在朦胧的灯光里，渗入一派月光的清辉，岸上垂柳淡淡的影子，在水里摇曳，犹如一幅画，令人陶醉。

At the entrance to a village or town or near a bustling market, a public square can often be found on a spacious water space for boats to berth and cruise. A bigger village or town may have a stage on the water, which can serve as a public venue for cultural and entertainment activities. The left picture shows a stage on the water in Suzhou. When a boat slides across the water, it leaves an elegant reflection and gentle ripples. In the silver moonlight and dim light penetrating through the windows of waterside houses, weeping willows sway elegantly on the bank and cast their shadows on the water, creating a captivating painting.

人们去农村或小镇中定期买卖货物的市场，在北方称赶集，川、黔等地区称赶场，湘、赣、闽、粤等地区称赶墟，新疆则称作赶巴扎。一般都在路边、桥头、村口等交通便利的地方设一固定场所，叫作集市广场。左图是浙北的一个集市广场。清晨，朴实憨厚的农民身着粗衣布裳，陆续不断地来到集市上，嘈杂声嗡嗡一片，热闹非凡。

在府邸庙观和大型民居前面，一般都有一个入口广场。主要是为了停放车轿和人流集散。外围建筑一般都是茶馆、酒楼、澡堂、店铺等。入口广场一般是为了突出建筑物的重要性，产生一些气势。我们知道人的视角大约在 60 度的范围内所见物体的形状才是清晰正常的，这个视域范围叫作正常视域。我们在街道上、庭院内，通常都不可能有这样长的距离去观察建筑。因此，入口广场就是观察建筑物立面的一个场所。建筑特性的不同，给入口广场带来的精神影响也不同，衙署宗祠广场的精神是威严、庄重，有一种装腔作势之感；而民居宅第的广场精神则是平易、活泼，充满人情味。每逢节日或有其他礼仪活动的日子，这里会演出地方戏、杂技等。

A permanent street market in a rural village or town is called "Ji" in northern China, "Chang" in places like Sichuan and Guizhou, "Xu" in places like Hunan, Jiangxi, Fujian, and Guangdong, and "bazaar" in Xinjiang Uygur Autonomous Region. Such markets are usually located on the roadside near a bridge or at the entrance to a village with convenient transport. These areas are called market squares. The left picture shows a market square in northern Zhejiang Province. Early in the morning, farmers in coarse clothes arrive at the market, making it quickly bustling with vendors and shoppers.

Typically, a square in front of a mansion, temple, or large residence is used to park vehicles and gather pedestrians. Around the square are buildings like teahouses, restaurants, bathhouses, and shops. An entrance square is often spacious to highlight the magnificence of the building complex behind it. Science has shown that humans' normal field of vision extends approximately 60 degrees, within which objects observed are clear and undistorted. However, in narrow lanes or courtyards, people are often unable to obtain such a wide field of vision, leaving the square at the entrance as the most ideal location for observing the facades of a building. The different functions of buildings bring different senses to the squares in front of their entrances. For instance, the square in front of the entrance to an old government building or ancestral temple usually looks majestic and solemn, with a pompous and histrionic sense. Those with a traditional residence are usually easygoing and lively, with a heavy breath of life. On festive occasions or during ceremonial events, performances of traditional operas, acrobatics, and other folk arts are often held in such squares.

上图是福建福州乡间某入口广场。建筑物很有特色，两边的封火墙使立面分为三段，尤其是一圈出檐，使建筑物之间有与邻为善的感觉。茶余饭后，人们可在廊下乘凉。那围栏是高度适宜的凳子，有人面向里面聊天说笑，有人面向外面抽烟凝思，乡土气息十分浓厚。

The picture above shows the square in front of a mansion in the countryside of Fuzhou, capital of Fujian Province. The buildings are distinct in terms of architectural style. The gables on both sides divide the facade of the building into three sections. In particular, the overhanging eaves extend to neighboring buildings, implying a sense of good-neighborliness. After dinner, neighbors chat with each other in roofed corridors to escape the summer heat. The balustrades are also designed to serve as seats, where some sit to chat and others smoke in meditation. All of this captures the leisure of rural life.

变幻的街景 Changing Streetscapes

交通干道在村镇的布置中具有划分区域的作用，因地形所限和其他人为的原因，不可能划分为完全方正整齐的平面。村镇中的主要街道只要地形许可则多取平直，尽量向棋盘式布局靠拢。一些次要的街巷往往曲曲折折，使街景逐渐展开，建筑依次出现，避免视线的一览无遗，产生一种不断变化的艺术效果。右图是江苏吴县盛泽的街景。

Alongside hosting traffic, streets also divide a village or town into blocks. For the confinement of topography or other reasons, few villages or towns can be divided into even blocks. If conditions allow, major streets are often as straight as possible, and the layout of the entire village or town adopts a chessboard-like style. Minor streets and lanes are usually zigzagging, creating a changing aesthetic effect to avoid an unblocked view of streetscapes and buildings. The right picture shows a street of Shengze Town, Wuxian County, Jiangsu Province.

尽量利用高大的建筑物，把它们组织到街道的对景中是常见的手法之一。如村镇中的钟鼓楼，跨街的牌楼、过街楼等。也有在丫形路口的尖端地段建造高大的庙宇以构成对景的。有的村镇把街做成“丁”字形或“之”字形，在交叉路口建造凸出的建筑物，避免一眼望穿。这里我们看到的是山西平遥的古城楼，在低矮的房屋中，它丰富了城镇的立体轮廓线。这是一座优秀的建筑，使人看了就能领略到一种威严。建筑物不仅是一个立面，更是外部结构与内部结构的有机综合体。这个综合体的每一个要素都要参与到整个艺术体验之中去。当我们站在远处看到这座市楼时，那立面丰满而富于节奏的轮廓线就像一幅画。接近看它时，建筑的雄伟气魄就逐渐变得显著起来，三层楼檐舒展遒劲，把结构的力量充分显示出来，走到楼的前面时，仰首上望，层层斗拱精雕细刻。市楼优雅而恬静、雄伟而壮观，有一种攫人的力量。这一切都是依照市楼建造者的安排纳入秩序的。这种复杂的和变化着的感受是建筑美属性中不可缺少的一部分。

It is common to use tall buildings, such as drum and bell towers, memorial archways, and cross-street structures, to form the highlights of streetscapes in old villages and towns. Some villages and towns built magnificent temples at Y-shaped crossroads to add dynamics to the streetscapes, and others purposely designed their streets T-shaped or Z-shaped and built high structures at the crossroads to avoid unblocked views. The picture depicts a gate tower in the ancient city of Pingyao in Shanxi Province. Setting off the lower residences, the tower enriches the skyline of the ancient city. The remarkable structure emits a majestic sense. A building is not just about appearance—it is a synthesis combining exterior and interior structures. Each element of the synthesis needs to participate in the effort to offer aesthetic experience. As we viewed the tower from a distance, its rich and dynamic outlines appeared like a painting. A close-up observation reveals its

unparalleled magnificence. The imposing protruding eaves of the three-story tower fully demonstrate the strength of its structure. Standing in front of the tower and looking up, one can see exquisitely carved *dougong*. Combining elegance and magnificence, the tower radiates inspiring power. All of it can be attributed to architects who incorporated this sense into the building when designing it. The complex and changing feeling is an undividable part of the aesthetic feature of the building.

磁器口在明清时期是重庆繁盛的江边市镇。沿街杂货店、屠户案桌、爆竹铺、成衣庄、理发店、布号、盐号，商贾云集，人声鼎沸，木楼瓦屋鳞次栉比。镇上店铺大多为前店后坊，楼上住人。除日常买卖以外，集市庙会更是繁盛，届时摊肆林立，百戏杂陈，农民身背竹篓，附近的人扶老携幼，人车熙攘，热闹非凡。右图是深幽的巷道，破旧的房屋，形成与热闹的市区迥然相异的宁静的居住环境。

Ciqikou was a bustling riverside town in Chongqing during the Ming and Qing dynasties. Its streets were lined with grocery shops, butcher's stalls, barbers, and crowded stores selling fireworks, apparel, and salt. There were rows upon rows of wooden buildings with tiled roofs. Many stores included a production area in the rear, while the second floor was used for bedrooms. Temple fairs would be even more bustling: In addition to performances of various folk arts, countless stalls and farmers hawking products carried on their backs transacted with shoppers old and young. The streets were packed with people and vehicles. The right picture shows a tranquil residential area with deep alleys and old houses which sharply contrasts the bustling downtown.

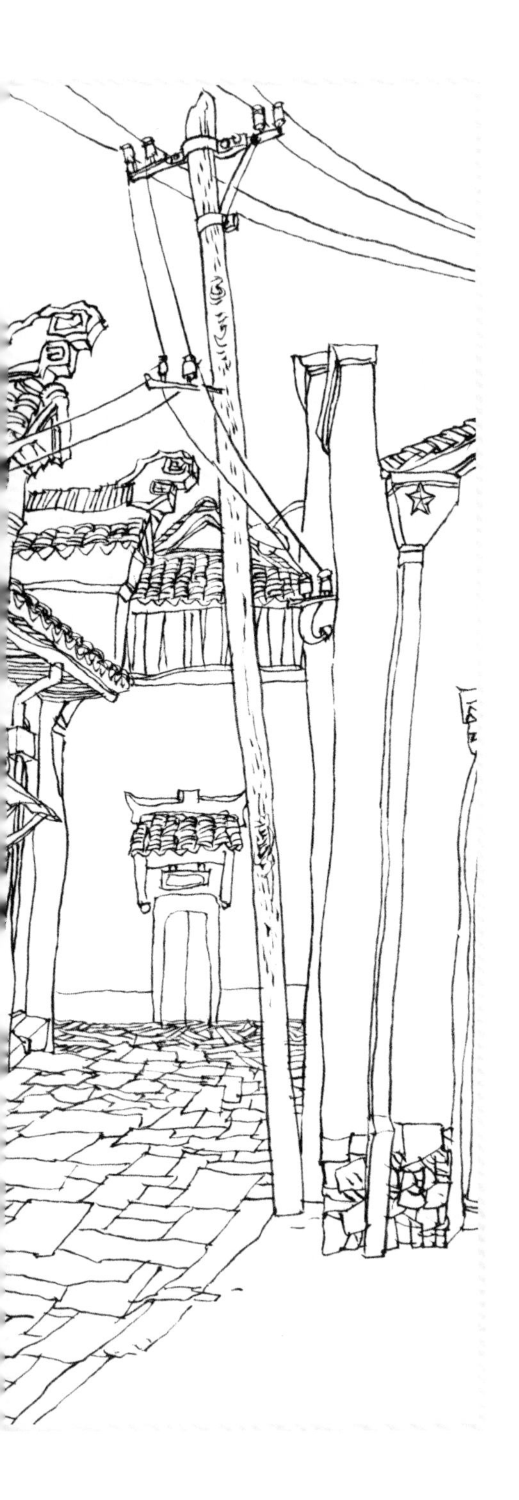

村镇的某些特有风貌是在其发展中自然形成的，但有些艺术面貌却是人们有意识地通过规划和建筑处理而成的。村镇中由水平方向展开的民居所组成的轮廓线一般是一条水平线，为了打破单调，常运用高耸的建筑物来丰富村镇的轮廓线。这些高耸建筑物突出于水平线之上，由它竖向的外形和一定的高度与水平线形成对比，增加了艺术上的趣味。这些高大建筑常常在城内或近郊山岗等制高点上，是最富有民族特点的风景点缀之一，也成为了一个村镇的标志和象征。左图我们看到的是江苏吴县光福镇，背景上有一座四方形的小塔，叫“香雪塔”。每年桂花开放时节，从离镇几千米的地方，徐风就会送来阵阵沁人心脾的清香，令人陶醉。

Some unique features of old villages and towns were formed naturally with the passage of time, but others can be attributed to deliberate planning and architectural design. To break the uniformity of the contour lines of residences in a village or town, some tall buildings were constructed. Their vertical heights contrast the almost unified horizontal heights of residences, creating an aesthetic touch. The tall buildings often stand in the center of the town or on nearby hills. They become the most typical scenic spots of ethnic flavor as the landmarks of villages and towns. The left picture shows Guangfu Town in Wuxian County, Jiangsu Province. The square-shaped pagoda in the background is the Xiangxue (Scented Snow) Pagoda. When laurels bloom, one can smell the refreshing fragrance blowing in the gentle breeze up to several kilometers away from the town.

06 牌坊 Memorial Archways

民居一般沿街修建，组成带形空间，在一些人们经常经过的地方常建有牌坊、券门，划分并丰富了空间。它们具有双重价值功能，既起到了入口的作用，也限定了区域。其意境体现在奇幻之中，奇源于幻就是这种建筑的寓意。牌楼还有点景作用，为普通的建筑赋予文化含意，构成人文景观。江苏徐州一牌楼，一边书写“五省通衢”，另一边是“大河前横”。它不仅给人以方位标识，使人想到黄河故道就在前面；而且重楼叠翠，令人遐思，奇中见幻，幻中见奇。当我们抬头去仰视那层层斗拱托起的飞檐铜铃时，奇与幻达到了有机统一，其艺术魅力令人陶醉。

Typically, residences line the streets, forming belt-like blocks. Memorial archways and arched doorways were built at locations with larger traffic flows to separate and enrich different spaces. They have dual functions: serving as entrances and defining the certain spaces. Such buildings inspire tourists' imagination. Moreover, memorial archways also convey cultural connotations, making them cultural attractions for sightseers. A memorial archway in Xuzhou City, Jiangsu Province, has a plaque reading "Thoroughfare to Five Provinces" on one side and "With a Big River Flowing in Front" on the other side. This not only indicates the city's prime geographical location, but also reminds people of the fact that the Yellow River once flew there. Behind it are towering buildings in the shade of trees, creating a dreamlike vista. As one looks up at its overhanging eaves supported by *dougong* and the bronze wind bells hung at the tips, the fascinating artistic charm of the memorial archway is fully manifested.

理髮處
華聲

07 河道 River Courses

江浙水网地区河道密如蛛网，水路交通纵横交织，池塘湖泊星罗棋布。在集镇村庄附近，人们利用天然有利条件开凿许多运河渠漕，把这些水面串联起来，构成一个密集的水路交通网络，因而水上运输十分发达。城市村镇沿河道发展，民居也沿河修建，与码头、驳岸等构成了江南特有的秀丽水乡风光。

The crisscrossing river courses in Jiangsu and Zhejiang provinces are like webs. There are also numerous ponds and lakes. In places near villages and towns, people dug canals to channel natural rivers and other water bodies, forming a developed water transport network. Villages and towns flourished along watercourses, and locals built their houses on riversides. Traditional residences, coupled with docks and revetments, constitute the unique scenery of water towns in southern China.

08 桥
Traditional Bridges

桥不仅解决了交通问题，而且在漫漫的历史长河中达到实用性与艺术性的结合，具有双重价值功能。古代的场镇选址与河流有着密切的联系，人们沿河修建住宅，跨河筑桥，使河流产生隔而不断的意蕴。桥与民居有着密切的联系。在南方的场镇中，桥往往被民居簇拥着，有的建筑倚桥而建。在右图右上部分我们可以看到，搭在桥上的台阶可以直接进入旁边一幢民居。在右图左下部分看到的是江南水乡券桥。从远处看去形成一个半圆景框，将错落有致的民居尽收眼底，圆拱倒映入水，形似一轮满月。为了通过大船或帆船，拱券常常做得很高。有时桥是一个镇的最高点，站在桥上可以俯瞰民居，颇为壮观。券的形状在南方多为半圆券，而北方多为双圆心券，尤其是略带椭圆状的三圆心券是我国民居形式的独到之处。这种券的做法是在发券时即底边用十分之一的半径再加半径的长度来左右画弧，到正中使用龙门石是又一个小圆心画弧。那弧券形的柔和生动是很富有画意的。苏州甪直镇的鸡鸭桥，桥拱入水部分仍呈圆弧，水上水下呈一圆圈，甚为独特。

More than just traffic facilities, bridges combine practicality with artistry over the long river of history, making them dually functional. In ancient times, people built residences along the rivers and bridges to connect the banks. Bridges and residences are closely related. In ancient towns of southern China, bridges are scattered amidst traditional residences, with some built against bridges. In the upper right of the right picture is a bridge with stairs directly leading to a nearby residence. In the lower left is an arched bridge common in water towns of southern China. From a distance, the bridge arch forms a semi-round “viewfinder” to enjoy nearby residences. Together with its reflection on the water, it forms a shape like a full moon. Bridge arches are usually

high so that big boats and sailboats can pass under. In some towns, bridges are the highest buildings and provide spectacular views of the rolling residences below. Most bridges in southern China feature a semi-round arch, and those in northern China two semi-round arches. Bridges with three semi-elliptical arches are unique to China. The radius of this kind of semi-elliptical arch is one-tenth longer than of a semi-round arch at the base, but the curve of the upper middle part is still like a round arch. The gentle curves of the semi-elliptical arches look lively and poetic. The Jiya Bridge in Luzhi Town, Suzhou City, remains a circular arc when it stretches into the water, forming a circle with its reflection.

券桥有石券桥和砖券桥，券孔以河宽而定，有单券、双孔，三、七、九、十一以至数十孔券不等。左下图是一座平、券结合桥。细雨纷纷，云烟缭绕，透过雨雾望去，那横跨河上的长桥忽隐忽现，恍若在虚空里。多么幽深神秘！

Arched bridges are often built with stone or brick. The span of the arches changes according to the width of the river. Alongside bridges with one arch, there are also bridges with multiple arches ranging from two to dozens. The lower left picture shows a plank bridge with arches. In the hazy drizzle, the long bridge spanning the river looks tranquil and mysterious, creating a fairyland-like scene.

平桥是最常用的桥，尤其是水浅的地方或只通小船的地方，平桥最好用，人过桥如履平地，不必费力。平桥有木桥、石桥或木石混合桥。桥在民居密集处，和民居融为一体，往往形成公共交往空间，有的桥上可以设茶座、集市。到了夏季，人们晚间在桥上乘凉，享受河上之清风。

However, flat bridges remain the most common. They are more practical, especially in places where the rivers are so shallow that only small boats can cruise. Pedestrians don't have to use as much energy to cross a flat bridge. Flat bridges are constructed with wood, stone, or mixes. Bridges are typically located in places with dense residences. They merge with the residential buildings and serve as venues for socializing. Some bridges even have tea stalls and markets. On summer nights, residents gather on bridges to enjoy the cool air from the rivers.

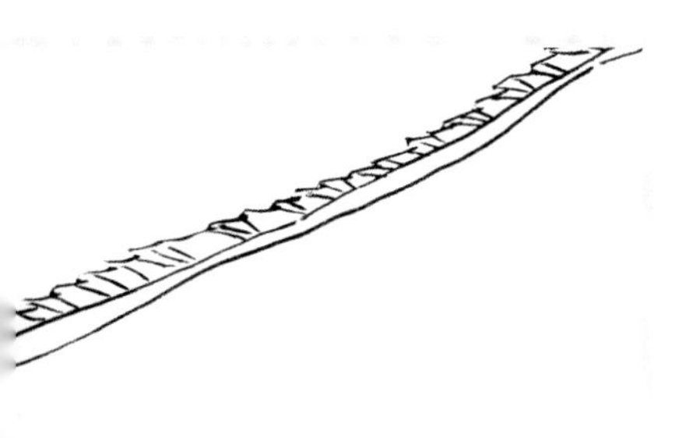

路棚
Roadside Booths

苏州自古就是我国最为繁华的城市之一，“络纬机杼之声，通宵彻夜”“四方商贾，蜂攒蚁集”。丝织业的发达，促进了农业和商业的繁荣，太湖的水产，洞庭东西山的茶叶、枇杷、柑橘名扬海内，富商官贾集居于此，修宅建园。东山是沿庭山脚弯曲延绵十里的一个小镇，七月盛夏，沿街走着，口干舌燥，弯曲的路上突然出现了个路棚，这真像是沙漠中的绿洲，来往的路人都在这儿憩息。凉风习习，实在惬意。路棚下面一边是茶馆，一边是食品店，生意兴隆。路棚的作用不单是为路人提供乘凉之处，更重要的是为了留住客人。

Suzhou has been one of China's most prosperous cities since ancient times. Historical records claim that "the sound of looms was heard all day and night" and "merchants from around the world swarmed the city." The booming silk weaving industry also led to the prosperity of agriculture and commerce. Aquatic products from Taihu Lake and the tea, loquat and orange from the Dongting East and West Hills are known across China. In the past, many wealthy merchants and high-ranking officials built their residences in the area. Dongshan is a small town that extends five kilometers along the foot of Dongting Mountain. In scorching July, roadside booths serving tea seem like oases after a long trek through the desert. It is pleasant to enjoy the cool air while sipping tea. Booths also serve other snacks. Such roadside booths are ideal resting places for walk-by customers.

私用码头 Private Quays

江浙水网地区交通和生活用水多依赖河渠，临水建筑很多。沿河行走，你会发现民居临水的一面多开后门，用条石砌筑踏步做成的私用码头，当地人称“河埠头”。私用码头和民居的关系安排大都巧妙精湛，结构合理，生趣盎然，往往不拘一格，别出心裁。和建筑的关系既有对比、衬映，又有协调、自然，使民居和河流产生联系，寓静止于流动之中，艺术语言非常明快。临水民居往往把厨房设置在沿河的一边，既可以看到洗濯、淘米的妇人忙忙碌碌，偶尔这家和那家的邻居打招呼，又可以看到船民摇橹沿河叫卖蔬菜鱼虾，家人外出或返回都在这里上下船只。

The water towns of Jiangsu and Zhejiang provinces use rivers for transport and domestic water. Most buildings are located along river banks. Many of the waterside residences have a backdoor facing the river and a private quay with stone stairs. Locals call this type of quay "Hebutou" ("river quays"). The private quays are skillfully connected to the residences. They feature reasonable structures, and each has a unique design that adds vitality to the riverfront. The quays establish harmony with surrounding buildings and link the motionless residences with the flowing river with sprightly artistic language. Riverside households usually place kitchens near the river. It is common for housewives to wash clothes and rice in the river while hospitably greeting passing neighbors. Boat operators sell vegetables and aquatic products like fish and shrimp along the river. Locals also use boats as the main means of transport when they go out.

祠堂公馆
Ancestral Temples and Guildhalls

民居中的公共性建筑物是指庙宇以外具公开性质的由居民集资修建的建筑，如公堂、会馆、桥梁等。桥梁在前面已专题叙述过。会馆是同一地区的人士，因常去外地经商为往返方便而在该地建造建筑以便歇脚、聚会、共叙友情、互通信息的，如山西会馆、四川会馆。会馆往往是一地区的驻外代表建筑，故一般都十分精美。在上面图片中，左上图是四川自贡西秦会馆，下图是湖南凤凰县戏台对面的看台。

In traditional Chinese architecture, public buildings are structures other than temples constructed with funds raised from the public such as memorial halls, guildhalls, and bridges. Guildhalls are buildings for merchants from the same place to rest, gather, socialize, and exchange information. The Shanxi Guildhall and the Sichuan Guildhall are pristine examples. As a region's representative offices in other parts of the country, guildhalls were often exquisitely built. The upper left picture shows the Xiqin Guildhall in Zigong City, Sichuan Province, and the lower picture shows a stage and opposite spectator stand in Fenghuang County, Hunan Province.

上图左上部分是福建厦门某祠堂。祠堂有两种，一种是供同族人共同祭祀祖先的，另一种是社会公众或某个阶层为共同祭祀某个人物而修建的。有的地方称祠堂为公堂。公堂是较大的村庄中居民共用的厅堂，如喜事、丧事等可在此举办典仪；居民可在此聚会或请艺人来此表演小型传统戏曲等。大型村庄或一般集镇都建有戏台等公众娱乐建筑，戏台的台面较高，各地风俗也不尽相同，一般高约七尺，有的更高，如上图右下部分云南大理的戏台。

The upper left of the picture above shows an ancestral temple in Xiamen City, Fujian Province. Essentially, there are two types of memorial temples. One is an ancestral temple for members of the same clan to enshrine and worship their ancestors, and the other is used to honor a celebrated figure venerated by the public or a certain spectrum of society. Ancestral temples are also called public halls in some places. A public hall is a magnificent building used by villagers to hold ceremonies such as weddings and funerals. It can also serve as a venue for gatherings and small-scale traditional opera performances. Typically, big villages or towns have public buildings with stages for entertainment purposes. Such stages vary in style in different areas. Most are about two meters tall, and a few are higher. The lower right of the picture above shows a stage in Dali, Yunnan Province.

12 公用空间 Public Spaces

横七竖八的竹竿跨越街道，晒着棉被、毛衣、裤衩……仿佛这不是一条街，而是居民住宅。只见一老汉扛着板凳大声吆喝着“磨剪子来抢菜刀”，呵，小镇的民风多么朴实，他们邻里关系和睦，彼此把公用街道作为自己家庭的院落，这是中国民居不同于现代公寓的主要特点之一。这里的居民最富人情味，住户普遍都把他们的厨房延伸到街缘上，街道成为居民公共活动空间，老人打麻将，小孩做游戏，老妪们一边唠家常，一边纳着鞋底，手里的针锥时而在头皮上擦两下，洋溢着浓厚的生活气息。

Bamboo sticks stretch from the balconies of residences to buildings on the opposite side of the street to dry quilts, sweaters, pants, and other clothes. In this way, the street becomes part of the residences. Carrying a stool on his back, an old man wanders down the street while shouting, “Any scissors or chopping knives to sharpen?” The neighbors in the small town live in harmony, and they make public streets part of their courtyards. This is a major difference between living in traditional residences and living in modern apartments. Local residents are hospitable, and many households feature kitchens right in the street. Roadsides are also public venues where the elderly play mahjong, children play games, and women chat while stitching soles of cloth shoes. The atmosphere of traditional living is strong.

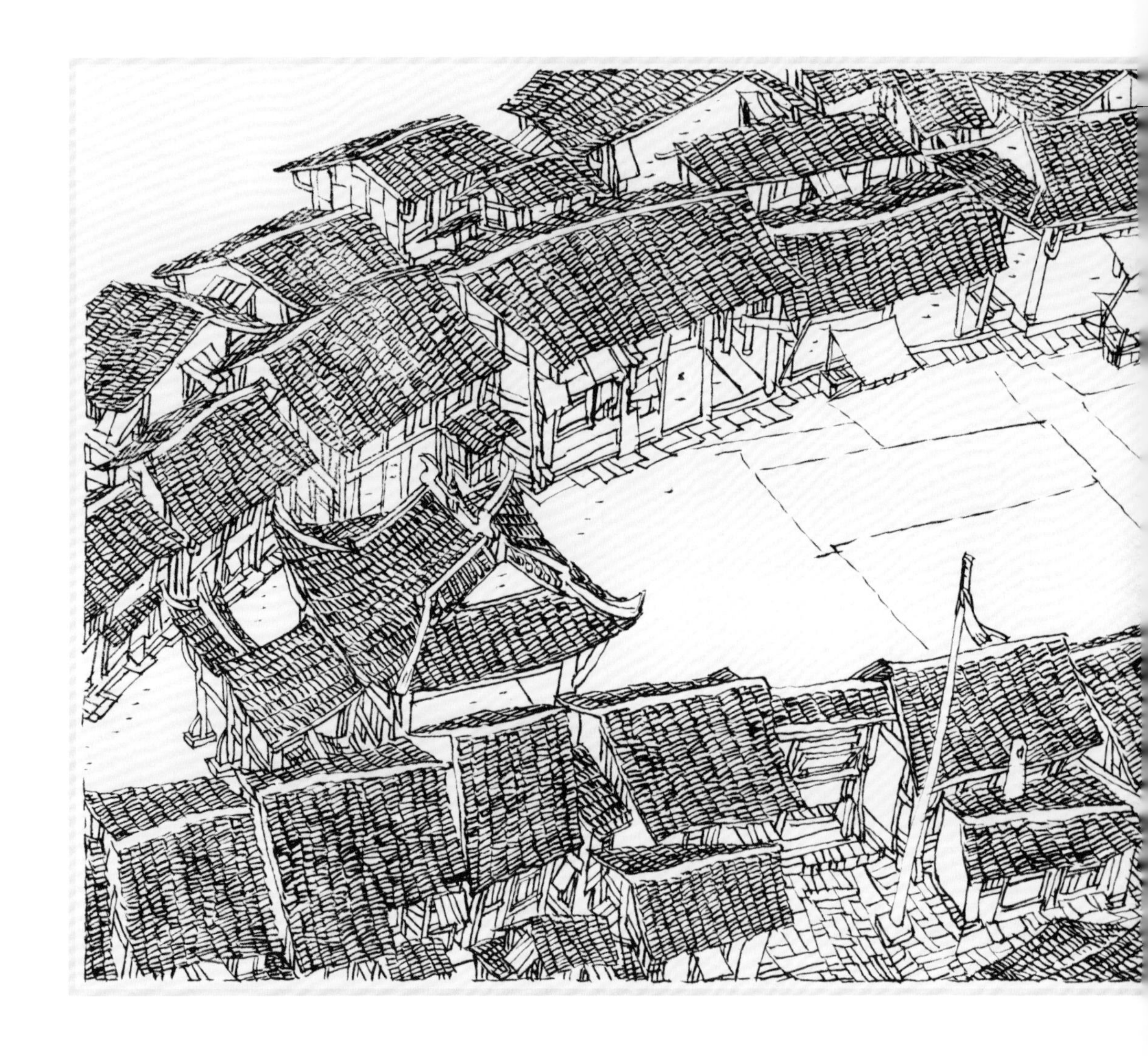

13 村镇举例——罗城古镇
Luocheng—A Good Example of Ancient Towns

在村镇的规划中，不乏优秀的例证，这里仅向大家介绍一下著名的罗城。罗城古镇在四川省乐山市犍为县境东北部，始建于明朝末年。整个集镇布局巧妙，建筑独特，一条南北走向的主街恰似一艘大船。中心是一个广场，既是集市广场，又是文化广场。广场两边是集市贸易的廊棚，中间是一座戏台，这种布局是商场设计中“两头吸收，中间消化”的典型范例。每逢节日，龙灯、狮子、高跷、旱船、珍马等花花绿绿，热闹非凡。这里我们看到的是罗城古镇示意图。古镇的建筑风格曾引起中外建筑学界的极大兴趣，澳大利亚的墨尔本市就仿照这一船形古镇修建了一座旅游新城。

Outstanding examples of planning of ancient villages and towns abound, of which Luocheng is one of the most famous. Built in the late Ming Dynasty, the ancient town of Luocheng is located in northeastern Qianwei County in Leshan City, Sichuan Province. The entire town boasts a unique layout. Its north-south main street resembles a huge boat, in the middle of which is a square serving as a market and a venue for cultural activities. On both sides of the square are roofed market stalls, and in the center is a stage. This layout is widely used in market design, with an aim to "attract customers from both ends and herd them through the middle." On festive occasions, a variety of folk art performances such as dragon lantern dancing, lion dancing, stilt walking, and boat dancing are staged in the square, attracting numerous visitors. This is a sketch of Luocheng Ancient Town. Its unique architectural style has drawn interest from architects around the world. A tourist town modeled after the boat-shaped layout of Luocheng was even built in Melbourne, Australia.

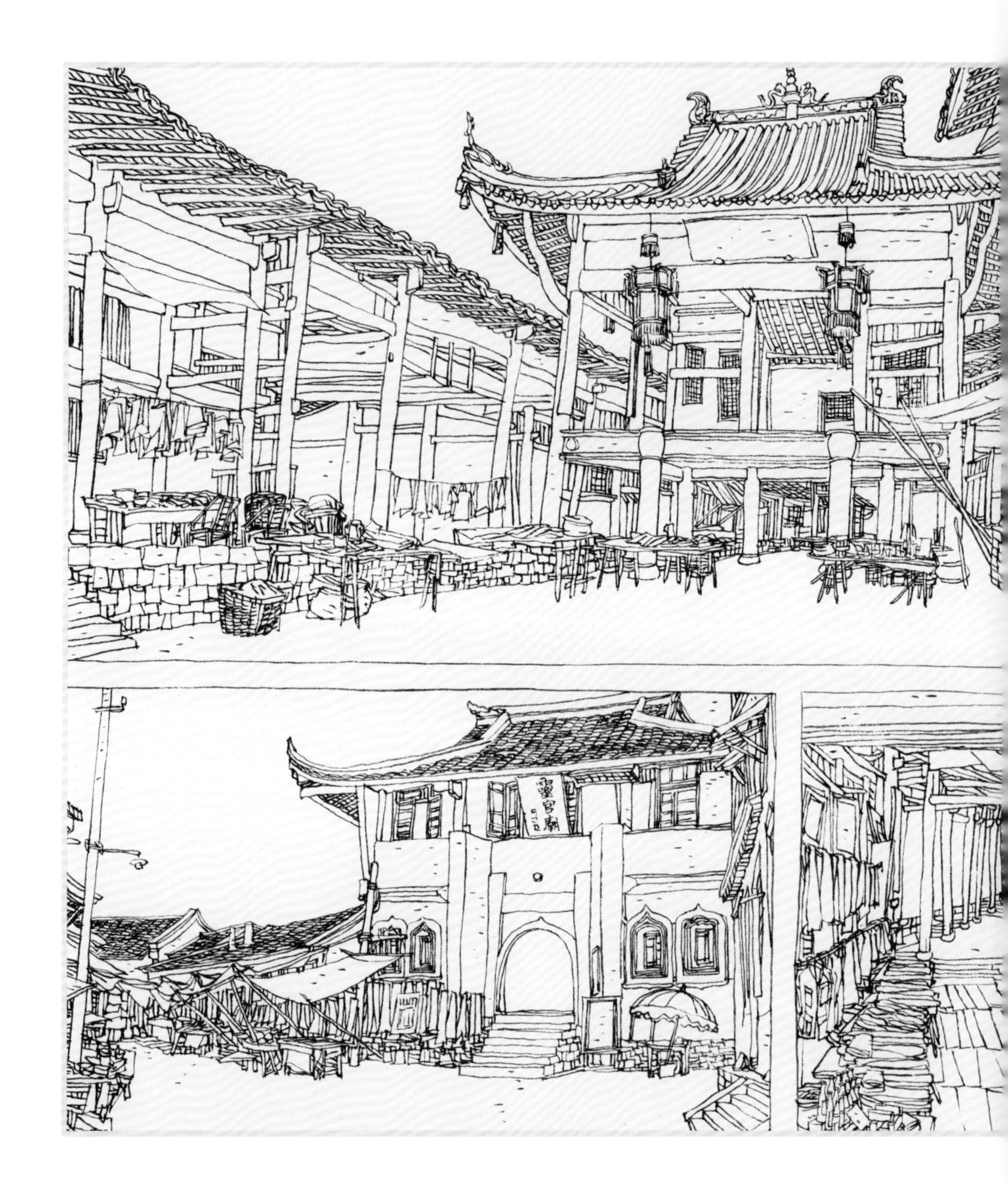
靈官廟

从这里（左侧）我们可以看到罗城古镇建筑的细部处理。

上图：“船中”的古朴戏台，两旁柱上的对联是“昆高胡弹灯曲绕黄梁，生旦净末丑功出梨园”。

下左图：灵官庙在“船尾”部分，犹似船的“尾篷”。这是一座清真寺，从楼下门窗的外形上可以领略到伊斯兰风格。

下右图：街道两旁长长伸展的棚廊，是无论阴晴雨雪都可以进行交易的市场。

These pictures (left) show details of some structures in the ancient town of Luocheng.

Above: The ancient stage in the middle of the “boat” and the pillars on each side carry a pair of couplets that mean, “The songs of all genres of Sichuan Opera lingers here, and all characters can perform on the stage.”

Lower left: The Lingguan Temple at the tail of the “boat” resembles the mat roofing. It is a mosque, and the door and windows on the first floor demonstrate a distinctive Islamic style.

Lower right: The roofed corridors extending on both sides of the street enable merchants to sell even on rainy and snowy days.

14 环境 Surrounding Environment

无论是小小的居室、深深的廊院、明亮的敞棚还是幽静的楼房，都离不开环境气氛的渲染，这种渲染是带有“强制性”的。民居之所以美，就是因为它们和当地的环境能够融为一体。藏族碉房必须放置在高原之上，蓝天白云的衬托给人以永恒的神秘之感。苏州的小楼必须是在狭窄曲折的深巷中，白墙黑瓦的相映给人以淡雅的神秘感。四川民居只有在青山翠竹的掩映中才能把我们的情感带进幽雅，情绪在舒畅凉爽的气氛中延伸。如果离开了环境的调节，民居的美也就不复存在了。

Exquisite residences, vast courtyards, bright corridors, and tranquil buildings cannot be divorced from their surrounding environment. Such a connection is “mandatory.” To a large extent, the beauty of traditional residences is their harmonious coexistence with the local environment. Only against a blue sky and white clouds over the plateau can Tibetan folk houses demonstrate an eternal mysterious sense. Only in zigzagging alleys can the elegance of traditional residences with white walls and black tiles in Suzhou be fully manifested. Only amidst green mountains and lush bamboo can traditional residences of Sichuan emit their pleasant and elegant ambience. The unique beauty of various types of traditional residences would disappear when they were removed from the local environment.

Traditional Chinese Residences

Written and Illustrated by Wang Qijun

First English Edition 2023

By China Pictorial Press Co., Ltd.

CHINA INTERNATIONAL COMMUNICATIONS GROUP

Address: 33 Chegongzhuang Xilu, Haidian District, Beijing, 100048, China

ISBN 978-7-5146-2078-8